Two Manuals Combined into One Book

Combat Breaching

and

FM 7-85 Ranger Unit Operations

Department of Defense

Pentagon
Publishing

Field Manual
No. 3-34.2

Headquarters
Department of the Army
Washington, DC, [pending date]

Combined-Arms Breaching Operations

Contents

Distribution Restriction: Approved for public release; distribution is unlimited.

---

*This publication supersedes Field Manuals (FMs) 5-101, 23 January 1985, and 90-13-1, 28 February 1991.

Page

Page

Preface

FM 3-34.2 provides combined-arms commanders, from company to brigade, and their staffs with the doctrine, tactics, and techniques needed to successfully overcome obstacles. It provides the commander with breaching fundamentals and information for planning, preparing, and executing breaching operations. This manual gives the coordinating and special staff officer a basis for synchronizing all combat multipliers during a breaching operation.

This is not a stand-alone manual. The user must have a fundamental understanding of the concepts outlined in FMs 5-71-2, 5-71-3, 5-71-100, 17-98, 20-32, 34-130, 71-1, 71-2, 71-3, 100-5, 100-7, 101-5, and 101-5-1. This manual also implements Standardization Agreement (STANAG) 2036.

Appendix A contains a metric conversion chart.

The proponent for this publication is HQ, TRADOC. Send comments and recommendations on Department of the Army (DA) Form 2028 directly to Commandant, United States (US) Army Engineer School, ATTN: ATSE-DOT-DD, Directorate of Training, 320 Engineer Loop, Suite 336, Fort Leonard Wood, Missouri 65473-8929.

Unless this publication states otherwise, masculine nouns and pronouns do not refer exclusively to men.

Chapter 1

Obstacle Breaching Theory

Breaching operations are conducted to allow maneuver despite the presence of obstacles. Obstacle breaching is the employment of a combination of tactics and techniques to advance an attacking force to the far side of an obstacle that is covered by fire. It is perhaps the single, most difficult combat task a force can encounter. Understanding breaching theory is the first step to understanding breaching tactics. Breaching is a synchronized combined-arms operation under the control of a maneuver commander. Breaching operations begin when friendly forces detect an obstacle and begin to apply the breaching fundamentals, and they end when battle handover has occurred between follow-on forces and a unit conducting the breaching operation. *Bulling through* or *forcing through* is not a breaching operation. Bulling through is a decision made when a commander must react immediately to extricate his force from an untenable position within an obstacle and no other breaching operations are possible. When a force is in a minefield receiving fires and taking heavy losses, the commander may decide to immediately bull through the minefield rather than withdraw or reduce the obstacle.

DEFINITIONS

1-1. The following definitions are inherent to obstacle breaching and are used throughout this manual:

- **Obstacle**. An obstacle is any obstruction that is designed or employed to disrupt, fix, turn, or block the movement of an opposing force (OPFOR) and to impose additional losses in personnel, time, and equipment on the OPFOR. Obstacles can exist naturally (existing), be man-made (reinforcing), or be a combination of both. A complex obstacle is a combination of different types of individual obstacles that requires more than one reduction technique (explosive, mechanical, manual) to create a lane through the obstacle. A reinforcing obstacle is an obstacle that is specifically constructed, emplaced, or detonated through military effort. (For more information on obstacle classification, see *FM 90-7*.) The types of reinforcing obstacles are—
  - **Tactical**. A tactical obstacle is employed to disrupt enemy formations, turn them into a desired area, fix them in position under direct and indirect fires, or block their penetration while multiplying the effects and capabilities of firepower.
  - **Protective**. A protective obstacle is employed to assist a unit in its local, close-in protection.

- **Lane**. A lane is a route through, over, or around an enemy or friendly obstacle that provides safe passage of a passing force. The route may be reduced and proofed as part of a breaching operation, be constructed as part of the obstacle, or be marked as a bypass.
- **Reduction**. This is a task to create and mark lanes through, over, or around an obstacle to allow the attacking force to accomplish its mission. Normally, engineers and reduction assets are used to reduce an obstacle.
- **Proofing**. Proofing verifies that a lane is free of mines and that the width and trafficability of the point of breach are suitable for the assault force. Proofing can be conducted visually (against surface-laid minefields), electronically (mine detectors), or mechanically (mine-clearing rollers [MCRs]). Proofing is conducted when the risk of live mines remaining in the lane exceeds the risk of loss (lives and equipment) to enemy fires while waiting to complete proofing. Some mines are resistant to some reduction assets; for example, magnetically fused mines may be resistant to a mine-clearing line charge (MICLIC). Proofing is vitally important, considering the vast variety of mines in use. It should always be planned for a breaching operation; however, the time available, the threat, or the mission may dictate that proofing not be done.
- **Clearing**. This is the total elimination or neutralization of an obstacle or a portion of the obstacle. Clearing operations are not conducted under fire and are usually performed by follow-on engineer forces.
- **Breach area**. The breach area is established and fully defined by the higher headquarters (HQ) of the unit conducting breaching operations. It is the area where a breaching operation occurs. The area must be large enough to allow the attacking unit to deploy its support force and extend far enough on the far side of the obstacle to allow follow-on forces to deploy before leaving the breach area. One technique is to establish the breach area using phase lines (PLs) or unit boundaries. The PL defining the far side of the breach area may be established as a battle handover line (BHL).
- **Farside objective**. The farside objective is the immediate objective of the attacking unit's assault force, and it may be oriented by the terrain or the enemy force. The higher HQ assigns the objective; however, the attacking unit normally subdivides the objective into smaller objectives to assign responsibilities and to control and focus the assault of subordinate forces. When breaching as part of a larger force, seizing the farside objective provides the necessary maneuver space for the higher unit's follow-on forces to move securely through the lanes, assemble or deploy, and continue the attack without enemy interference.
- **Bypass**. A bypass is a tactical task that involves maneuvering around an obstacle, a position, or an enemy force to maintain the momentum of advance. When a unit bypasses an obstacle, it physically changes the direction of movement to avoid the obstacle. This must be done with caution because it might play into the enemy's hand. Bypassed obstacles and enemy forces are reported to higher HQ.

- **Point of penetration**. This is the location, which is identified on the ground, where the commander concentrates his efforts at the enemy's weakest point to seize a foothold on the farside objective. This is achieved along a narrow front through maneuver and direct and indirect fires that are accurately placed against enemy forces. A commander conducting a breaching operation establishes a point of penetration that supports planning locations for the reduction area and the seizure of the farside objective.
- **Reduction area**. This is a number of adjacent points of breach that are under the control of the breaching commander. The commander conducting the attack determines the size and location of the reduction area that supports the seizure of a point of penetration (see *Figure 1-1*).
- **Point of breach**. This is the location at an obstacle where the creation of a lane is being attempted. Initially, points of breach are planned locations only. Normally, the breach force determines the actual points of breach during the breaching operation.

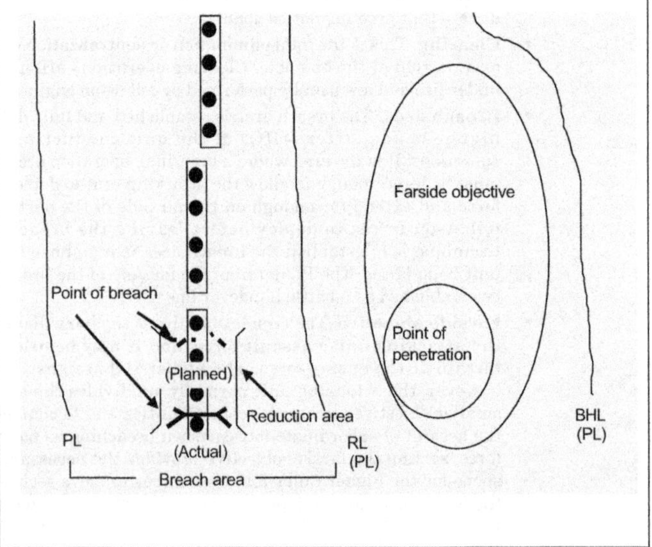

Figure 1-1. Breach Area

BREACHING TENETS

1-2. Successful breaching operations are characterized by applying breaching tenets. These tenets should be applied whenever an obstacle is encountered in the area of operations (AO), whether during an attack or during a route clearance operation. The tenets are—

- Intelligence.
- Breaching fundamentals.
- Breaching organization.
- Mass.
- Synchronization.

INTELLIGENCE

1-3. Success depends largely on the force commander's ability to *see the battle space*. He must identify how the enemy is using the terrain to minimize the risk of surprise. This is particularly true when attempting to counter the enemy's use of obstacles. The force commander does this with the intelligence preparation of the battlefield (IPB) process. During the IPB process, the situation template (SITEMP) is developed. The SITEMP is a graphic depiction of expected threat dispositions based on threat doctrine and the effects of the battle space for a particular course of action (COA). Based on the commander's estimates, the staff develops a SITEMP for multiple enemy COAs. The SITEMP helps drive the maneuver planning in support of war gaming and the development of the event template. The SITEMP also drives the decisions the commander must be prepared to make. The commander uses the decision-support template (DST) to convey these decisions in graphic form. The IPB process helps achieve success by focusing available intelligence-gathering assets toward answering the commander's priority intelligence requirements (PIR). Intelligence that is gathered later by recon becomes the foundation for developing and revising a SITEMP. (See *Appendix B* for information on threat mine operations.)

1-4. Time-constrained planning requires the rapid development of the enemy's SITEMP. A minimal SITEMP, complete with templated obstacles, allows the development of an initial event template and provides initial PIR for the intelligence-collection plan (ICP). At a minimum, the unit can template an obstacle trace that is precise, as far as the location, and represents the total expected enemy countermobility effort but does not depict individual obstacles. The staff plots the actual composition of the trace based on incoming obstacle reports. Adequate time for intelligence collection is critical in developing an accurate picture of the battle space, yet time is normally in short supply. While templating is essential to focus the ICP, too much time can be spent on templating at the expense of intelligence, surveillance, and reconnaissance (ISR).

1-5. In any operation where enemy obstacles can interfere with friendly maneuver, obstacle intelligence (OBSTINTEL) is always one of the information requirements (IR) and should become PIR. Finding enemy obstacles or seeing enemy obstacle activity validates and refines the Intelligence Officer's (US Army) (S2's) picture of the battle space.

OBSTINTEL helps the S2 determine the enemy's intentions, plans, and defense strength. The force engineer is the unit's expert on enemy countermobility capabilities. He assists the S2 in templating enemy obstacles, analyzing OBSTINTEL, and analyzing the effects of terrain.

1-6. An unverified enemy template can lead to disaster because the force may aim an attack at the wrong place. Units may deploy to reduce expected obstacles early, wasting mission time to *feel* their way into nonexistent obstacles; or they may blunder into an unexpected obstacle or an enemy engagement area (EA).

1-7. OBSTINTEL can be a critical indicator to verify an enemy template. The force's Operations and Training Officer (US Army) (S3), S2, and engineer establish effective OBSTINTEL collection by determining specific obstacle IR. Examples of information that is needed to fulfill obstacle IR include—

- The location of existing and reinforcing obstacles.
- The orientation and depth of obstacles.
- Conditions of the soil (in the case of a minefield) to determine the ability to use tank plows.
- The presence, location, and type of wire.
- Lanes and bypasses.
- The composition of the minefield (buried or surface-laid antitank [AT] and antipersonnel [AP] mines, antihandling devices [AHDs], and the depth of the mines).
- Types of mines and fuses.
- The location of enemy indirect-fire systems that can fire into the breach area.
- The composition of complex obstacles.
- Areas between successive obstacle belts.

1-8. OBSTINTEL is particularly important for discovering the types of mines and mine fuses the enemy has employed. The engineer depends on this information since he must determine which reduction techniques offer the best chance for success and minimize the risk to the breach force. This may require a dismounted recon patrol to examine mines within a minefield. The enemy may employ security patrols, have forces overwatching the obstacle, or place forces within the obstacle to prevent obstacle recon.

1-9. Obtaining OBSTINTEL requires dedicated collection assets. These assets identify survivability positions and obstacle emplacement activity. Aviation units, who are tasked to perform recon in the area, may be able to provide valuable OBSTINTEL information. Ground-surveillance radar may also provide useful information.

1-10. Recon forces should have the capability to collect OBSTINTEL. To be successful in this collection, the obstacle recon efforts must be tied into the maneuver force's overall reconnaissance and surveillance (R&S) plan. A force ensures this capability by training dedicated recon assets on collecting required obstacle information or by augmenting recon forces with engineer recon teams provided by the supporting engineer force. An engineer recon team moves with scouts or a patrol and conducts dismounted recon of templated or confirmed obstacles. This team has the flexibility to dismount and develop the details of the obstacle. Personnel engaged in a recon mission for OBSTINTEL should be used rarely, if ever, to reduce obstacles during the mission; although they make ideal guides to breach forces for subsequent breaching operations. Inadvertent detonation during reduction may compromise engineers and scouts, defeating the recon mission. It may also compromise the entire attack. *FM 5-170* describes capabilities, limitations, and employment concepts of engineer recon teams.

BREACHING FUNDAMENTALS

1-11. Suppress, obscure, secure, reduce, and assault (SOSRA) are the breaching fundamentals that must be applied to ensure success when breaching against a defending enemy. These fundamentals will always apply, but they may vary based on the specific battle-space situation (mission, enemy, terrain, troops, time available, and civilian considerations [METT-TC]).

Suppress

1-12. Suppression is a tactical task used to employ direct or indirect fires or an electronic attack on enemy personnel, weapons, or equipment to prevent or degrade enemy fires and observation of friendly forces. The purpose of suppression during breaching operations is to protect forces reducing and maneuvering through an obstacle.

1-13. Effective suppression is a mission-critical task performed during any breaching operation. Suppressive fires in sufficient volume serve to secure the reduction area. Successful suppression generally triggers the rest of the actions at the obstacle. Fire control measures ensure that all fires are synchronized with other actions at the obstacle. Although suppressing the enemy overwatching the obstacle is the mission of the support force, the breach force should be able to provide additional suppression against an enemy that the support force cannot effectively suppress.

Obscure

1-14. Obscuration must be employed to protect forces conducting obstacle reduction and the passage of assault forces. Obscuration hampers enemy observation and target acquisition and conceals friendly activities and movement. Obscuration smoke deployed on or near the enemy's position minimizes its vision. Screening smoke employed between the reduction area and the enemy conceals movement and reduction activities. It also degrades enemy ground and aerial observations. Obscuration must be carefully planned to provide maximum degradation of enemy observation and fires, but it must not significantly degrade friendly fires and control.

Secure

1-15. Friendly forces secure the reduction area to prevent the enemy from interfering with obstacle reduction and the passage of the assault force through the lanes created during the reduction. Security must be effective against outposts and fighting positions near the obstacle and against overwatching units, as necessary. The far side of the obstacle must be secured by fires or be occupied before attempting any effort to reduce the obstacle. The attacking unit's higher HQ has the responsibility to isolate the breach area by fixing adjacent units, attacking enemy reserves in depth, and providing counterfire support.

1-16. Identifying the extent of the enemy's defenses is critical before selecting the appropriate technique to secure the point of breach. If the enemy controls the point of breach and cannot be adequately suppressed, the force must secure the point of breach before it can reduce the obstacle.

1-17. The breach force must be resourced with enough maneuver assets to provide local security against the forces that the support force cannot sufficiently engage. Elements within the breach force that secure the reduction area may also be used to suppress the enemy once reduction is complete.

Reduce

1-18. Reduction is the creation of lanes through or over an obstacle to allow an attacking force to pass. The number and width of lanes created varies with the enemy situation, the assault force's size and composition, and the scheme of maneuver. The lanes must allow the assault force to rapidly pass through the obstacle. The breach force will reduce, proof (if required), mark, and report lane locations and the lane-marking method to higher HQ. Follow-on units will further reduce or clear the obstacle when required. Reduction cannot be accomplished until effective suppression and obscuration are in place, the obstacle has been identified, and the point of breach is secure. Obstacle-reduction techniques are discussed in more detail in *Appendix C*.

Assault

1-19. A breaching operation is not complete until—

- Friendly forces have assaulted to destroy the enemy on the far side of the obstacle that is capable of placing or observing direct and indirect fires on the reduction area.
- Battle handover with follow-on forces has occurred, unless no battle handover is planned.

BREACHING ORGANIZATION

1-20. A commander organizes friendly forces to accomplish the breaching fundamentals quickly and effectively. This requires him to organize support, breach, and assault forces with the necessary assets to accomplish their roles (see *Table 1-1*, page 1-8).

Table 1-1. Relationship Between Breaching Organization and Breaching Fundamentals

Breaching Organization	Breaching Fundamentals	Responsibilities
Support force	Suppress Obscure	Suppress enemy direct-fire systems covering the reduction area. Control obscuring smoke. Prevent enemy forces from repositioning or counterattacking to place direct fires on the breach force.
Breach force	Suppress (provides additional suppression) Obscure (provides additional obscuration in the reduction area) Secure (provides local security) Reduce	Create and mark the necessary lanes in an obstacle. Secure the nearside and far side of an obstacle. Defeat forces that can place immediate direct fires on the reduction area. Report the lane status/location.
Assault force	Assault Suppress (if necessary)	Destroy the enemy on the far side of an obstacle that is capable of placing direct fires on the reduction area. Assist the support force with suppression if the enemy is not effectively suppressed. Be prepared to breach follow-on and/or protective obstacles after passing through the reduction area.

Support Force

1-21. The support force's primary responsibility is to eliminate the enemy's ability to interfere with a breaching operation. It must—

- Isolate the reduction area with fires.
- Mass and control direct and indirect fires to suppress the enemy and to neutralize any weapons that are able to bring fires on the breach force.
- Control obscuring smoke to prevent enemy-observed direct and indirect fires.

1-22. Suppression is critical for a successful breaching operation; therefore, allocation of adequate forces to the support force is essential. The commander must allocate enough direct- and indirect-fire systems to the support force to allow it to accomplish its missions. When resourcing the support force, consider the losses it will receive as it fights its way into its support-by-fire (SBF) position. To increase the survivability of the support force, the commander may establish critical friendly zones (CFZs) in support of the support force once it has occupied its SBF positions. A CFZ is an area, usually a friendly unit or location, that the maneuver commander designates as critical to protect an asset whose loss would seriously jeopardize the mission. Covered by a radar sector, the CFZ supports counterfire operations by providing the most responsive submission of targets to the fire-support system when rounds impact inside the CFZ. (For more detailed information on CFZs, see *FM 6-121*.)

1-23. The support force must seize an SBF position, seeking maximum protection from the terrain. If possible, the support force should follow a covered or concealed route to the SBF position, take up its assigned sectors of fire and observation, and begin to engage the enemy. It may have to make some adjustments to its direct-fire plan if the situation dictates. Observation is particularly critical. Artillery observers with the support force may initially bring indirect fires on enemy positions to fix and suppress the enemy. The support force adjusts the artillery-delivered obscuring smoke, if necessary, to protect the breach and assault forces as they approach the reduction area.

Breach Force

1-24. The breach force assists in the passage of the assault force by creating, proofing (if necessary), and marking lanes. (For more information on the initial lane-marking pattern, see *Chapter 3*.) The breach force is a combined-arms force. It includes engineers, reduction assets, and enough maneuver forces to provide additional suppression and local security. The breach force may apply portions of the following breaching fundamentals as it reduces an obstacle:

- **Suppress**. The breach force must be allocated enough maneuver forces to provide additional suppression against various threats, including—
  - Enemy direct-fire systems that cannot be effectively observed and suppressed by the support force due to the terrain or the masking of the support force's fires by the breach force as it moves forward to reduce the obstacle.
  - Counterattacking and/or repositioning forces that cannot be engaged by the support force.
- **Obscure**. The breach force employs vehicle-mounted smoke systems and smoke pots, if necessary, for self-defense and to cover lanes while the assault force is passing.
- **Secure**. The breach force secures itself from dismounted threat forces who are providing close-in protection of the obstacle. The breach force also secures the lanes through the tactical obstacles, once they are created, to allow safe passage of the assault force.
- **Reduce**. The breach force performs its primary mission of reducing the obstacle. To support the development of a plan to reduce the obstacle, the composition of the obstacle system must be an IR.
- **Assault**. The breach force assaults through the point of breach to the far side of an obstacle and seizes the reduction area.

1-25. The breach force is comprised of two subordinate elements—security and reduction:

- **Security element**. The security element is comprised primarily of maneuver forces. It is normally responsible for providing additional suppression, obscuration, and local security.
- **Reduction element**. The reduction element is comprised primarily of reduction assets (mine-clearing blades [MCBs], MICLICs, armored vehicle-launched bridges [AVLBs], armored combat earthmovers

[ACEs], dismounted engineers, and so forth). It is normally responsible for reducing the obstacle.

Chapter 2 provides more details on the roles, responsibilities, and command and control (C^2) of the security and reduction elements.

1-26. The breach force must be able to deploy and begin reducing the obstacle as soon as enemy fires have been suppressed and effective obscuration is achieved. It can expect enemy artillery fires within a matter of minutes. CFZs should be activated at the point of breach before the commitment of the breach force.

1-27. Engineers within the breach force are allocated with the reduction assets necessary to clear mines, nonexplosive obstacles, and small gaps. Scouts, engineers, or other recon assets are also capable of finding local bypasses or existing lanes in the obstacle system.

1-28. The commander allocates engineers and reduction assets based on the number of lanes required, the terrain, and the type of obstacle. The breach force must be able to create a minimum of one lane for each assaulting company or two lanes for the assaulting task force (TF).

1-29. The commander should plan for a 50 percent loss of mobility assets in close combat. As a planning factor, he should allocate one engineer platoon with the necessary reduction assets to create and mark one lane.

1-30. After the breach force has reduced the obstacle and passed the assault force through, the breach force may be required to hand over the lane to follow-on units. At a minimum, the lanes must be marked and their locations and conditions reported to higher HQ and follow-on units as prescribed in the unit's standing operating procedures (SOP).

Assault Force

1-31. The assault force's primary mission is to destroy the enemy and seize terrain on the far side of the obstacle to prevent the enemy from placing direct fires on the created lanes. The assault force may be tasked to assist the support force with suppression while the breach force reduces the obstacle.

1-32. The assault force must be sufficient in size to seize the point of penetration. Combat power is allocated to the assault force to achieve a minimum 3:1 ratio on the point of penetration. In a breaching operation, particularly TF and above, the assault force normally maneuvers as a separate force attacking through a reduced obstacle. However, breach and assault assets may maneuver as a single force when conducting lower-level breaching operations, such as an independent company team conducting an attack.

1-33. If the obstacle is defended by a small enemy force, assault and breach forces' missions may be combined. This simplifies C^2 and provides more immediate combat power for security and suppression.

1-34. Fire control measures are essential, since support and breach forces may be firing on the enemy when the assault force is committed. Suppression of overwatching enemy positions must continue and other enemy forces must remain fixed by fires until the enemy has been destroyed. The assault force must assume control for direct fires on the assault objective as support and breach forces' fires are lifted or shifted.

MASS

1-35. Breaching is conducted by rapidly applying concentrated efforts at a point to reduce the obstacle and penetrate the defense. Massed combat power is directed against the enemy's weakness. The location selected for breaching depends largely on the weakness in the enemy's defense, where its covering fires are minimized. If friendly forces cannot find a natural weakness, they create one by fixing the majority of the enemy force and isolating a small portion of it for attack.

Support Force

1-36. The support force suppresses the enemy to eliminate effective fires on the breach force. Smoke and terrain are used to help isolate the force under attack. Suppression requires the commander to mass enough overwatching direct fires to protect the breach force.

1-37. Normally, the TF isolates and destroys platoon-size elements in succession, starting with the platoon identified as the easiest to overwhelm. In *Figure 1-2*, the TF commander employs a tank company team to fix two defending platoons and uses screening smoke (target AA1102) to help isolate the platoon selected for attack. Objective (OBJ) Cowboy has been identified by the TF commander as the point of penetration. The center and eastern platoons are suppressed by direct and indirect fires from a second tank company team.

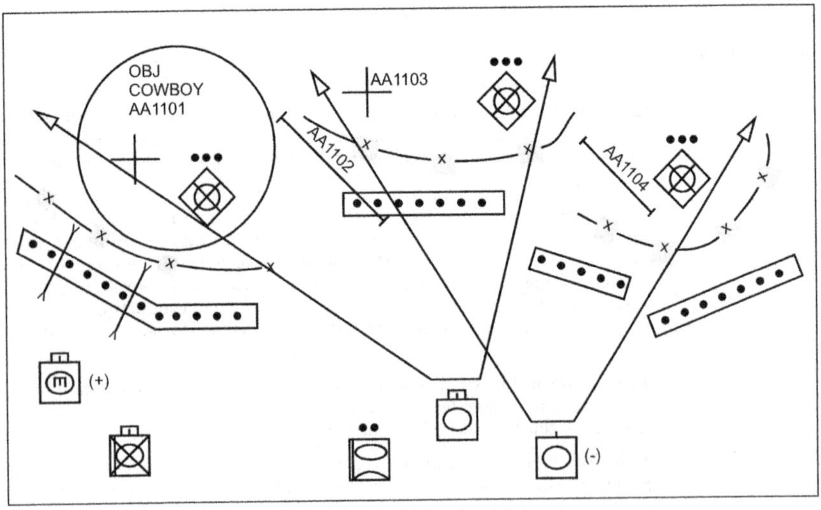

Figure 1-2. Technique for Achieving Mass in the Breach

1-38. The support force should be provided with reduction assets to reduce the impact that unexpected obstacles or scatterable minefields have on their approach to and occupation of SBF positions. Failure to do so may greatly impact on synchronizing the entire breaching operation. As a technique, a unit may create a mobility reserve that supports the main effort throughout the operation. Initially, the mobility reserve may support the support force until it has seized its SBF positions and then shift its support to the breach or assault force.

Breach Force

1-39. The commander also masses engineers and reduction assets to reduce the obstacle. The breach force is organized and equipped to use several different reduction techniques in case the primary technique fails; for example, if a key vehicle is destroyed or if casualties render dismounted engineers ineffective. Additional reduction assets are present to handle the unexpected. Normally, 50 percent more reduction assets than required for obstacle reduction are positioned with the breach force.

1-40. The breach force masses reduction efforts against the obstacle to ensure that it will successfully create enough lanes. A mounted TF-size assault force requires at least two lanes, but more lanes will speed passage through the danger area. The breach force attempts to simultaneously create as many lanes as necessary to ensure the rapid passage of the assault force through the obstacle system. This may require the breach force to simultaneously use a combination of mechanical, explosive, and manual techniques. For example, an engineer platoon may use two MICLICs as the primary technique to reduce a minefield. As soon as the engineer platoon fires the second MICLIC, engineer squads employ manual explosive techniques to create three additional lanes through the minefield. The commander must carefully plan and synchronize the creation of additional lanes to reduce the potential for fratricide with assaulting troops. The distance between the lanes depends on the enemy, the terrain, the need to minimize the effects of enemy artillery, the direct-fire plan of the support force, C^2, and reduction-site congestion.

Assault Force

1-41. Achieving necessary mass for the assault requires the breach force to open enough lanes through the obstacle to permit rapid passage and the buildup of forces on the farside. The size of the assault force determines the number of lanes initially created. A mounted TF requires one lane per assaulting company team. A dismounted assault force requires one lane for each leading assault platoon. The tactical situation may require additional lanes to quickly pass a large assault force through the obstacle to achieve sufficient combat-power ratio.

1-42. The principle of mass influences the selection of the point of breach; the task organization of the support, breach, and assault forces; and the integration of engineers in movement or attack formations. *Figure 1-2*, page 1-11, shows a technique to achieve mass in the breach.

1-43. The need to generate enough mass and the maneuver space available strongly influence which echelon can conduct the breaching operation. A company team cannot simultaneously mass sufficient fires, reduce the obstacle, and assault the defending position unless it is a simple obstacle defended by no more than one platoon. A TF should have a combat-power ratio of 3:1 to reduce an obstacle defended by a company, and a TF is normally the echelon used to execute a breach. A brigade has sufficient combat power to attack a complex, well-defended obstacle; but a brigade has difficulty deploying all its combat power within range. Normally, a brigade breaches by isolating a small segment of the defense (platoon or company) that the TF can then attack as the breaching echelon. If obstacles and defenses are in-depth, mass is achieved by passing additional TFs through the obstacle to continue the attack.

SYNCHRONIZATION

1-44. Breaching operations require precise synchronization of the breaching fundamentals by support, breach, and assault forces. Failure to synchronize effective suppression and obscuration with obstacle reduction and assault can result in rapid, devastating losses of friendly troops in the obstacle or the enemy's EA.

1-45. A combined-arms breach is a complex operation by nature. Support, breach, and assault forces must apply the SOSRA breaching fundamentals within a short time and distance. The support force masses its direct fires and controls indirect fires in concert with breach and assault forces' maneuvers. The commander must employ smoke at the right time and place to maximize its effectiveness or risk hampering his own target acquisition and C^2. The breach force must have the right reduction asset for the type of obstacle encountered. Engineers must ensure that they do not prematurely exhaust reduction assets needed to reduce subsequent obstacles. *Table 1-2*, page 1-14, shows the complexity of a combined-arms breach.

1-46. The commander ensures synchronization through proper planning and force preparation. Fundamentals to achieve synchronization are—

- Detailed reverse planning.
- Clear subunit instructions.
- Effective C^2.
- Well-rehearsed forces.

Detailed Reverse Planning

1-47. Synchronizing a combined-arms breach begins by using the reverse-planning process to ensure that actions at obstacles support actions on the objective. Planning the breach without regard to actions on the objective leads to disaster. During COA development, the staff analyzes the relative combat power and compares enemy and friendly strengths and weaknesses. The staff then generates options based on the commander's guidance and arrays the forces necessary to accomplish the mission. The staff develops the options by analyzing the objective and allocating forces to accomplish the mission. The commander decides how he must attack the objective to accomplish his mission. This decision drives where, how, and with what force he must support, breach, and assault through the enemy's obstacles. The commander approves the scheme of maneuver for the breaching phase of the operation

Table 1-2. Breach Complexity

Action	Element	Time (Minutes)	Controlled By
Develop the situation (verify the boundary of the enemy obstacle system).	Force in contact	M to 2	S3
Maneuver the support force into the overwatch position.	Support	M + 2 to 15	Support cdr
Maneuver the assault force into the covered assault position.	Assault	M + 2 to 15	Assault cdr
Call for artillery.	DS artillery	M + 2 to 15	FSO
Build smoke.	Mortars	M + 5 to 10	FSO
Suppress the enemy with direct fires.	Support	M + 15 to 29	Support cdr
Suppress the enemy with artillery fires.	DS artillery	M + 10 to 29	FSO
Maintain smoke.	DS artillery/mortars	M + 10 to 30	FSO
Maneuver the breach force to the breach location.	Breach	M + 20 to 23	Reduction cdr
Reduce the obstacle, and prepare two lanes.	Breach	M + 23 to 30	Engineer ldr
Place smoke pots.	Breach	M + 23 to EOM	Reduction cdr
Shift direct fires off the objective.	Support	M + 29 to 30	Assault cdr
Shift indirect fires beyond the objective.	DS artillery	M + 29 to 30	Assault cdr
Assault to destroy the enemy on the far side of the obstacle.	Assault	M + 30 to 45	Assault cdr
Reorganize to continue the mission.	TF	M + 45 to EOM	S3
NOTE: M = Contact with the obstacle			

that achieves adequate suppression, obscuration, and security. He also creates sufficient lanes to rapidly project combat power on the point of penetration and the far side of the objective, not just to the far side of the obstacle. Reverse planning gives purpose to the breaching plan that supports mission accomplishment.

1-48. The commander begins developing COAs as discussed in *FM 101-5*. He determines relative combat-power ratios, generates options, and arrays forces to accomplish the mission. He then develops the scheme of maneuver, including actions on the objective. In *Figure 1-3*, a three-company-team TF is conducting a deliberate attack to destroy a motorized infantry company (MIC) in the defense. The commander believes seizing OBJ Cat to be decisive. This is the point from which reverse planning of actions on the objective begins. The commander uses Team C to seize the initial foothold (OBJ Bird) (point of penetration) into the enemy's defense. The focus of breach planning is to get Team C to its objective. The number and location of lanes are driven by Team C's maneuver on OBJ Bird as the assault force. The commander now plans how the breach force must maneuver to reduce the obstacle. Supporting fires from Company A and Team B in SBFs 15 and 25 are planned to support the reduction of the obstacle and the assault on the objective. Reverse planning continues to drive the maneuver formation to ensure that forces are in the correct relative positions to accomplish their missions.

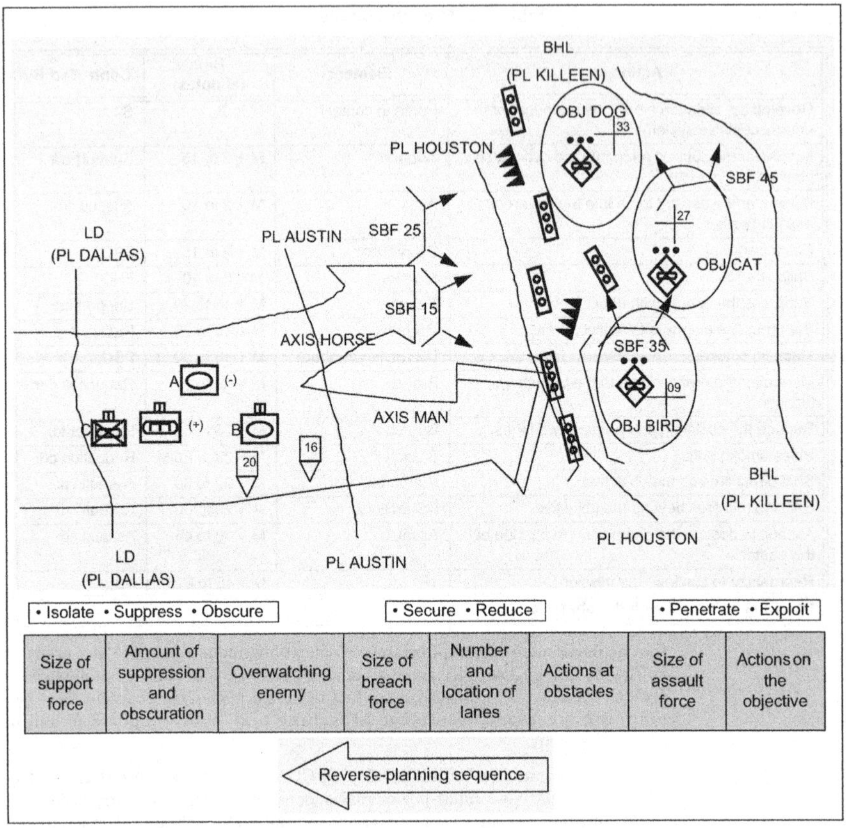

Figure 1-3. Reverse Planning

Clear Subunit Instructions

1-49. Subordinate units must clearly understand their missions, how their missions relate to the phases of the operation, and what role adjacent units play in the overall plan. The commander may use an execution matrix to synchronize his forces. An execution matrix is a superb synchronization tool, and it lists subunit instructions sequentially in relation to key events or the sequence of the attack. It also provides subordinate commanders with an understanding of how their missions and those of adjacent units fit into the overall plan. More importantly, it allows subordinates to better track the battle and coordinate their own maneuver with that of adjacent units. This is critical to achieving unity of effort between support, breach, and assault forces. *Figure 1-4* shows a sample execution matrix for the scheme of maneuver in *Figure 1-3,* page 1-15.

<table>
<tr><th></th><th>Company A (Support Force)</th><th>Team B (Support/ Assault Force)</th><th>Team C (Assault Force)</th><th>Team Engineer (Breach Force)</th><th></th></tr>
<tr><td rowspan="3">Units</td><td></td><td></td><td>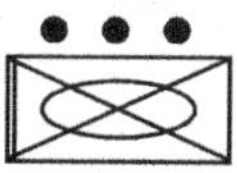</td><td></td><td rowspan="3">Force Allocation</td></tr>
<tr><td></td><td>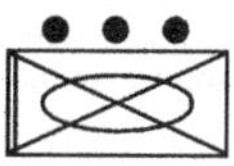</td><td>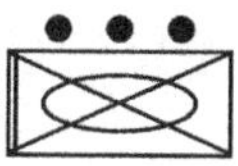</td><td></td></tr>
<tr><td></td><td></td><td></td><td></td></tr>
<tr><td>LD to PL Austin</td><td>Move to left of Team B; echelon left.</td><td>Lead TF.</td><td>Trail team engineer.</td><td>Trail Team B center.</td><td rowspan="3">Maneuver formation</td></tr>
<tr><td>PL Austin to PL Houston</td><td>Attack with Team B on Axis Horse.
Occupy SBF 25.</td><td>Lead attack on Axis Horse.
Occupy SBF 15.</td><td>Occupy CP 20.</td><td>Lead attack on Axis Man.</td></tr>
<tr><td>Obstacle reduction</td><td>SBF 25.
Orient TRP 09.
On order TRP 27.</td><td>SBF 15.
Orient TRP 09.</td><td>Occupy CP 16.
Identify lane.</td><td>Create and mark two lanes (vicinity AB123456).</td></tr>
<tr><td>Attack on OBJ Bird</td><td>Shift fires to TRP 33.</td><td>Shift fires to TRP 27.</td><td>On order, move through lane to seize OBJ Bird.</td><td>Assist passage of Team C.
Secure lanes.</td><td rowspan="2">Reduce</td></tr>
<tr><td>Attack on OBJ Cat</td><td>Orient TRP 33.</td><td>On order, lift fires and occupy CP 16.
Identify lanes.</td><td>On order, seize OBJ Cat.</td><td>Secure/improve lanes.</td></tr>
<tr><td>Attack on OBJ Dog</td><td>Lift fires.
Consolidate on SBF 25.</td><td>On order, seize OBJ Dog.</td><td>Occupy SBF 45.
Orient TRP 33.
On order, lift fires.</td><td>Assist passage of Team B.
Secure/improve lanes.</td><td rowspan="2">Action on OBJ</td></tr>
<tr><td>Consolidate</td><td>SBF 25.
Orient north.</td><td>OBJ Dog.
Orient northeast.</td><td>OBJ Cat.
Orient east.</td><td>Support hasty defense operations.</td></tr>
</table>

Figure 1-4. Sample Execution Matrix

Effective C^2

1-50. Effective C^2 is paramount to mission success. C^2 is integrated into the plan through the use of maneuver and fire control measures and the positioning of key leaders to see the battle space. Maneuver control measures enable the commander to graphically convey his intent, scheme of maneuver, and subunit instructions. Relating subunit actions to the terrain is critical to successful execution.

1-51. In *Figure 1-3*, page 1-15, the commander uses PLs to initiate the breaching operation and to control actions on the objective. At PL Austin, the support force maneuvers to occupy SBFs 15 and 25. The assault force occupies checkpoint (CP) 16 to prevent premature advancement to the obstacle, reduction-site congestion, and the engagement of forces before lanes are created and marked. The commander uses target reference points (TRPs) on obvious terrain features to orient, focus, and shift suppressive direct and indirect fires.

1-52. Key leaders must be able to see the battle space to make informed decisions. Nowhere is this more true than during the breaching operation. The commander must position himself where he can best control the battle. Since effective suppression is the most critical event during breaching operations, the commander may position himself with the support force. This enables him to personally influence fire control and facilitate the necessary cross talk between breach and assault forces. The S3 may initially move with the breach force to track the progress of obstacle reduction and anticipate the commitment of the assault force. A commander who feels that his personal influence is required with the breach or assault force must make a conscious effort to track the entire battle and must not focus completely on the reduction of the obstacle or the assault itself.

Well-Rehearsed Forces

1-53. The most effective synchronization tool available to the commander is the rehearsal. The inherent complexity of the breaching operation makes rehearsals at every level essential to success. The commander must give his subordinates time to plan how they will execute their assigned missions and time to rehearse the plan with their unit. The TF facilitates rehearsals by preparing rehearsal sites and managing their use. The TF commander must be sensitive to the impact TF-level rehearsals have on subordinate planning and preparation, and he must carefully choose the time and place of the rehearsal. Company teams rehearse their support, breach, and assault roles. TF rehearsals focus on synchronizing the maneuver of support, breach, and assault forces to achieve the breaching fundamentals. Rehearsals also highlight key events that must be coordinated during breach execution.

Typical Breaching Operations

1-54. *Table 1-3*, page 1-18, shows the types of breaching operations versus the sizes of the enemy forces overwatching the obstacle.

Table 1-3. Types of Breaching Operations Versus Enemy Sizes

Enemy Overwatching the Obstacle	Maneuver Units Conducting Breaching Operations		
	Brigade	TF	Company
MIBN	X		
MIC	O	X	
MIP		O	X
NOTES: X = Normal (combat-power ratio 3:1) O = Desirable (combat-power ratio >3:1)			

BREACHING OPERATIONS IN SUPPORT OF DELIBERATE AND HASTY ATTACKS

1-55. Attacks take place along a continuum based on the knowledge of the enemy's capabilities and disposition and the intentions and details of the friendly force's planning and preparation. The terms *deliberate attack* and *hasty attack* refer to the opposite ends of that continuum and describe characteristics of the attack:

- **Deliberate attack**. Friendly forces have more knowledge of the enemy situation and have more time to plan and prepare for the enemy. With this knowledge, friendly forces conducting the deliberate attack can develop detailed plans, task-organize to accomplish the mission, and execute rehearsals down to the lowest levels.
- **Hasty attack**. The enemy situation is vague, or it changes rapidly. Friendly forces have little or no time to plan or prepare for this type of operation. Because the primary purpose of the hasty attack is to maintain momentum or take advantage of the enemy situation, it is normally conducted only with the resources that are immediately available.

1-56. Breaching operations may be required to support an attack anywhere along the continuum. During a deliberate attack, friendly forces, such as engineers and smoke platoons, are task-organized before the attack to best support the concept of the operation. In a hasty attack, there may be insufficient time to task-organize assets, such as engineers and reduction assets, in the best manner. Additionally, the unit may not have time to plan the breaching operation in detail but may have to rely on the unit's SOP for execution. Regardless of where the attack falls along the continuum, the breaching tenets apply when conducting the breaching operation in support of the attack.

PLANNING SEQUENCE

1-57. The scheme of maneuver, engineer operations, fires, air defense, and actions at obstacles are all based on the same SITEMP. Planners develop the breaching plan using the following sequence:

- Reverse planning begins with actions on the objective.

- Actions on the objective drive the size and composition of the assault force.
- The size of the assault force determines the number and location of lanes to be created.
- Lane requirements and the type of obstacle drive the amount and type of mobility assets task-organized to the breach force.
- The ability of the enemy to interfere with the reduction of the obstacle determines the size and composition of the security element in the breach force.
- The ability of the enemy to mass fires on the point of breach determines the amount of suppression required and the size and composition of the support force.

Chapter 2

Breaching Operations

For a breaching operation, a unit develops a scheme of maneuver specifically designed to cross an obstacle and continue a mission. Maneuver company teams, TFs, and brigades can conduct breaching operations. Normally, a TF executes a breaching operation and the company teams are assigned as support, breach, and assault forces. Units will conduct a breaching operation when—

- The force-allocation ratio indicates that a confirmed enemy situation is beyond the subordinate unit's capabilities to successfully conduct a breaching operation. (When companies cannot successfully conduct a breaching operation, the battalion will.)
- A subordinate unit has failed in its attempt to breach an enemy's tactical obstacles.

NOTES:
1. See *Appendix D* for a breaching scenario and special planning considerations for breaching operations in restricted terrain.
2. See *Appendix E* for information on route clearance operations.

PLANNING

2-1. Units develop plans to conduct a breaching operation using the following military decision-making process (MDMP):

Step 1. Receipt of mission.
Step 2. Mission analysis.
Step 3. COA development.
Step 4. COA analysis (war game).
Step 5. COA comparison.
Step 6. COA approval.
Step 7. Orders production.

The following paragraphs highlight key portions of the MDMP as it applies to the planning of a breaching operation. For a complete discussion of the MDMP, see *FM 101-5*.

RECEIPT OF MISSION

2-2. Before receiving a higher HQ order, a unit should receive a series of warning orders (WOs) and available intelligence products. If the products are not issued with the WOs, the higher HQ liaison officer (LO) should locate and deliver the products as soon as they become available. Intelligence products are essential for the unit to begin developing its own intelligence products and to facilitate parallel planning.

2-3. Upon receiving the higher HQ SITEMP, the unit develops standard terrain products using TerraBase II or a similar computer-software program. (For examples of TerraBase products, see *FM 5-71-3*.) This set of terrain products is called a terrain-visualization mission folder. (For more information on this subject, see *FM 5-105*.) Terrain products that support the planning of a breaching operation should be included in the unit's SOP and may include—

- Visible area plots from each of the templated enemy locations to the expected EA.
- Visible area plots from potential SBF positions and points of breach to the templated enemy location.
- Oblique view of the AO.
- Perspective view of intervisibility lines.
- Visible area plots from named areas of interest (NAIs), when NAIs are developed.

2-4. Visible area plots from each of the templated enemy locations help determine if the enemy would select these locations. For example, the enemy would not select a location directly behind a small hill because the hill would prevent the enemy from firing into a kill sack. Additionally, by combining the visible area plots on an overlay, the enemy's ability to mass fires is more easily understood. It also helps select SBF locations, obscuration locations, and sectors of fires and helps determine the breach organization for future operations.

2-5. Visible area plots from potential SBF positions and points of breach assist future planning by allowing the commander to visualize the support and breach forces' ability to suppress the enemy. Multiple plots along the length of an SBF position may be necessary to fully understand the suitability of an SBF position.

2-6. An oblique view of the AO helps the commander better visualize the terrain. A perspective view of intervisibility lines in the AO allows him to visualize how the unit can best use the terrain during a breaching operation. It also provides an idea of how the enemy may use the intervisibility lines when positioning its assets and special munitions.

2-7. Visible area plots from developed NAIs can be used to provide positioning guidance for observation posts (OPs). These products are issued to recon assets who are tasked to provide information on various NAIs.

2-8. When receiving a mission, pay particular attention to the actions the higher HQ takes in support of the breaching operation or its decisions that influence the breaching operation. Responsibilities of the higher HQ include—

- Resourcing the unit conducting the breaching operation with additional or special assets not normally task-organized to subordinate units.
- Fixing enemy forces to prevent them from repositioning and interfering in the breaching operation.

- Using scatterable mines (SCATMINEs), deep fires, attack helicopters, and close air support (CAS) to isolate the farside objective and prevent a successful enemy counterattack.
- Planning a forward passage of lines by follow-on forces through the unit conducting the breaching operation. This must include a plan to upgrade/maintain lanes through the obstacle and to create additional lanes, if necessary, for the passage of follow-on forces.
- Providing criteria to transition to its own breaching operation.

MISSION ANALYSIS

2-9. During mission analysis, the commander conducts the IPB. A product of the IPB process is the SITEMP. The SITEMP is a graphic depiction of expected threat disposition if it adopts a particular COA. The staff must combine higher HQ intelligence with its knowledge of the enemy to produce a SITEMP with the necessary detail to successfully plan a breaching operation. A SITEMP should include—

- Likely enemy EAs.
- The location and orientation of enemy forces.
- Counterattack objectives and the location of enemy reserve elements.
- The location and range of all direct- and indirect-fire systems (tanks, AT weapons, artillery).
- Enemy obstacle systems, including tactical and protective obstacles and SCATMINEs. The template must depict the enemy's countermobility capability.
- The enemy's use of nuclear, biological, chemical (NBC) weapons, including the ranges of delivery systems.
- The location of enemy target-acquisition assets.
- Likely air avenues of approach (AAs), including fixed and rotary-wing aircraft.
- Positioning of enemy air-defense artillery (ADA) assets.

2-10. Selecting the initial commander's critical-information requirements (CCIR) is a part of mission analysis, and it is crucial to begin planning recon operations that will result in useful intelligence, including OBSTINTEL. Typical CCIR for a breaching operation may include the—

- Location, composition, and orientation of the obstacle and available bypasses in the vicinity of the reduction area and the point of penetration (PIR).
- Location and composition of enemy forces that are capable of employing direct and indirect fires on the point of breach (PIR).
- Maintenance status and location of all reduction assets (friendly forces information requirements [FFIR]).
- Status of commitment criteria for the breach force (PIR and FFIR).

NOTE: The initial CCIR is developed during COA analysis.

COURSE-OF-ACTION DEVELOPMENT

2-11. When developing a COA, the commander must remember that the breaching operation is only a portion of the higher HQ operation. The unit must ensure that the end state of the COA supports the higher HQ concept of the operation. As stated in *FM 101-5*, one of the criteria that each developed COA must meet is completeness. In a breaching operation, units use detailed reverse planning, develop a plan to achieve the breaching fundamentals, develop clear subunit instructions to develop a complete COA, and begin to synchronize the operation.

2-12. The detailed reverse-planning process, the SITEMP, and the terrain products produced during mission analysis are used to determine the size and composition of support, breach, and assault forces. The reverse-planning process begins with actions on the objective and includes templated enemy obstacles.

2-13. The assault force's task and purpose drive its composition. The assault force's mission is to seize terrain on the farside of the obstacle to prevent the enemy from placing direct fires on created lanes. Consider the following when determining the assault force's size and composition:

- Higher HQ requirements in terms of seizing or securing objectives.
- The terrain.
- The expected disposition, composition, and strength of the enemy when the assault force is committed.

2-14. The assault force's size determines the number and location of lanes to be created. The breach force creates and marks lanes through or around obstacles to support the rapid passage of the assault force. An assaulting company team requires a minimum of one lane, and an assaulting TF requires at least two lanes.

2-15. Lane requirements and the type of obstacle drive the amount and type of reduction assets task-organized to the breach force. As stated in *Chapter 1*, the commander should plan for the loss of about 50 percent of his reduction assets during a breaching operation. To ensure that the breach force is adequately resourced, the initial allocation of assets is continually reviewed as additional intelligence information (including OBSTINTEL) is received.

2-16. The enemy's ability to interfere with obstacle reduction determines the size and composition of the security element in the breach force. The security element's size and composition within the breach force can also be determined through the—

- Terrain analysis and the terrain's impact on the support force's ability to suppress the enemy.
- Amount of terrain the breach force will occupy at the obstacle during reduction and the extent that the breach force will mask the fires of the support force.
- Analysis of the enemy's expected reaction to the reduction effort.

2-17. Detailed terrain analysis and the enemy's expected size, disposition, and composition determine the enemy's ability to mass fires at the point of breach. That ability, in turn, determines the amount of suppression that is required and the support force's size and composition. Expected losses that the support force will incur from the line of departure (LD) to the SBF position due to maintenance and enemy air, chemical, and artillery attacks should be considered. Additionally, analyzing the enemy's equipment type, range, and lethality can indicate the expected losses that the support force will incur while in an SBF position.

2-18. When developing COAs for a breaching operation, consider the following:

- **Suppress**.
  - Identify the support force's axis of advance to SBF positions. The support force's mobility requirements must be identified based on the SITEMP and its axis of advance to SBF positions.
  - Establish SBF positions for each element in the support force.
  - Establish TRPs to control the support and breach forces' direct fires.
  - Determine the location and size of CFZs to protect the support force.
  - Plan essential indirect-fire targets to assist in suppressing the enemy.
  - Develop a plan for lifting or shifting direct and indirect fires.
  - Deploy the support force into a position to fire simultaneously on the enemy from several directions, if possible.
  - Attack by fire before or during obstacle reduction to suppress an overwatching enemy. If circumstances allow, forces may bypass the obstacle and assault the overwatching enemy before reduction. If the assault successfully eliminates the enemy, it also eliminates direct fires on the breach force. If the assault does not succeed in destroying the enemy, it fixes them and reduces the enemy's ability to place fires on the breach force. Direct- and indirect-fire suppression from the support force is necessary to support this assault.
  - Use fires to suppress the enemy's indirect fires. The enemy's indirect fires are extremely dangerous during a breaching operation since the breach force may be exposed for a lengthy period. Friendly artillery and air assets support the breaching operation by attacking enemy overwatch positions, mobile reserves, and artillery positions that are capable of firing on the point of breach. Friendly forces jam enemy fire control nets. The fire-support officer (FSO) designates points of breach as CFZs for a target-acquisition battery so that enemy artillery is treated as a priority target for counterbattery support.
- **Obscure**.
  - Determine the location of smoke target(s).
  - Identify terrain features that mask the force's point of breach, and limit the number of enemy forces that can effectively observe and fire on the breach force, thus aiding in obscuration. Other natural, limited-visibility conditions (darkness, fog) should be exploited if possible. Smoke is generally the primary obscurant; however, it is

a double-edged sword because it may attract enemy attention and degrade friendly target acquisition and C^2.

- Determine the quantity, location, and type of obscurant required to screen breach and assault forces and obscure enemy fires. This is particularly important for breaching complex obstacles where the breach force is exposed for an extended time. Smoke is important in deceiving the enemy as to the breaching location and interfering with enemy fires. If smoke platoons are available, synchronize them with artillery-delivered and organic smoke.
- Determine smoke requirements by using *FM 3-50*. Blanket, haze, or curtain-screening smoke degrades enemy observation of assault and breach forces. A smoke platoon is capable of providing support for a 0.5- to 1.4-kilometer front. Friendly forces use a combination of carefully placed smoke pots and indirect fires to augment smoke support. Howitzers and mortars supporting friendly forces can build and maintain smoke, but smoke effectiveness depends on the number and rate of smoke rounds delivered. Using indirect fire to provide smoke competes with the friendly force's high-explosive (HE) and dual-purpose improved conventional munition (DPICM) missions against an enemy. A commander must decide the priority for the use of indirect fire, and his decision must be clear to subordinates. Vehicles equipped with smoke-grenade launchers can provide local smoke for reduction operations and for troop passage during an assault. However, these systems are self-defense mechanisms and friendly forces must not rely on them to supply all obscuration for a breaching operation.

- **Secure**.
  - Plan direct-fire control measures as necessary.
  - Use terrain and the positioning of nearside and farside security SBF near the point of breach to facilitate the survivability of the breach force.
  - Plan a CFZ and ADA coverage around the reduction area for survivability and security of the breach force.
  - Plan local security at the breach site to suppress overwatching enemy forces that are close to the obstacle.
- **Reduce**.
  - Identify primary and alternate reduction techniques.
  - Determine the tentative location of lanes and the separation between lanes.
  - Determine the breach force's task organization.
  - Position the breach force, and identify positions and routes that it uses as it maneuvers toward the obstacle to minimize its impact on hindering the support force's suppressive fires.
- **Assault**.
  - Plan to pass the assault force through lanes created in the obstacle.
  - Develop objectives for the assault force.

- Develop SBF positions to be occupied by the assault force to facilitate passage of follow-on forces.
- Plan direct- and indirect-fire control measures that support the seizure of the farside objective.
- Identify all mobility requirements for the COA, including whether the assault force requires engineer support for reducing protective obstacles and fortifications within the defensive position.

2-19. Each element within the breach organization must completely understand which tasks it is responsible for in breaching operations. Additionally, each COA should identify which force is the main effort throughout the breaching operation. For example, the COA should define when the main effort shifts from the support force to the assault force. The following are the different types of forces and their tasks and responsibilities:

- **Support force**.
  - Suppress enemy elements capable of placing direct fires on the point of breach to protect the breach force.
  - Suppress enemy elements capable of placing direct fires on the assault force. This may require the adjustment of SBF positions.
  - Call for and adjust indirect fires, including obscuration. A combat observation lasing team (COLT) or Striker will be required for Copperhead munitions.
  - Fix enemy forces that are capable of repositioning.
- **Breach force**.
  - Reduce lanes in the obstacle.
  - Provide local security (farside and nearside).
  - Provide additional suppression of enemy forces overwatching the obstacle.
  - Mark and report the location of created lanes.
  - Assist the passage of the assault force through created lanes.
- **Assault force**.
  - Seize the farside objective.
  - Reduce protective obstacles.
  - Prevent enemy direct fire from interfering with follow-on forces as they pass through lanes.
  - Provide clear routes to the BHL for follow-on forces.
  - Conduct battle handover with follow-on forces.

2-20. The assault force's mission is unique. It must reduce the enemy's protective obstacles, assault the enemy's position, and secure terrain that is essential to the passage of follow-on forces. The assault-force commander must thoroughly understand the higher HQ scheme of maneuver. The following information is particularly important:

- Objectives the assault force must seize.
- The end state, in terms of SBF positions, which the assault force must occupy so that follow-on forces can continue the attack.

- The number of lanes the breach force plans to create to facilitate the passage of the assault force.
- The higher HQ plan to provide suppressive fires and/or obscuring smoke.
- Planned points of breach and the point of penetration.
- Lane requirements (number and lane-marking type) to support the passage of follow-on forces.
- The sequence of follow-on units to be passed through lanes and the criteria to begin passing follow-on units through lanes that the assault force created.

2-21. Typically, the enemy emplaces protective obstacles 50 to 500 meters in front of its positions and between its forward platoons and companies. These obstacles are designed to prevent a mounted assault and to fix or break up a dismounted assault. Close-in obstacles may be a combination of wire, AP and AT mines, fortifications, and entrenchments. All obstacles are covered by direct fire. Mines may be surface-laid or buried and well-camouflaged. Mines can be deployed with AHDs that are command-detonated or trip-wire-activated.

2-22. The assault force commander considers the following breaching fundamentals when developing his plan:

- **Suppress**. If a company team is the assault force as part of a TF breaching operation, other company teams within the TF will normally have the responsibility to suppress the enemy from designated SBF positions on the objective area. These teams may be the same support force that protected the breach force; however, terrain factors may require repositioning to provide effective support for the assault force.
- **Obscure**. Mortars and vehicle-mounted obscurants are primary means of obscuration for the assault force. It may also employ hand-held smoke grenades and smoke pots to provide additional obscuration. If available, smoke may be employed away from the point of breach to confuse enemy forces as to the disposition of the assault force.
- **Secure**. Local security must be established before conducting reduction operations on protective obstacles.
- **Reduce**.
  - The assault force's reduction efforts center on creating dismounted/mounted lanes in protective obstacles, reducing fortifications, and widening lanes to support follow-on forces. The commander should task-organize the assault force with sufficient reduction assets to create a minimum of one lane per assaulting platoon. There may be insufficient time to reorganize engineers and reduction assets to the assault force after crossing the LD. Therefore, engineers and reduction assets supporting the assault force are task-organized to the maneuver force they will support.
  - During the planning process, the assault force commander must decide whether he will assault mounted or dismounted. This decision impacts the reduction methods that will be used. A

mounted assault allows the use of tank plows and MICLICs. During a dismounted assault, friendly forces rely on manual obstacle reduction, which is very time-consuming. The composition of follow-on forces must also be considered. This may require widening initial lanes that the assault force created, especially when a dismounted assault is followed by mechanized follow-on forces.

- The assault force normally decentralizes C^2 for the plan's execution. For this reason, an assault force commander must allocate engineers, reduction assets, and demolitions to squads and platoons to provide reduction capability throughout his force.

- **Assault.** Below-ground operations require clearing the enemy's trench lines and bunker complexes. Dismounted infantry and engineers, in concert with tanks and Bradley fighting vehicles (BFVs), conduct these operations. These actions must be planned in detail and rehearsed to ensure proper execution. See *FMs 71-1* and *7-10* for more information.

COURSE-OF-ACTION ANALYSIS (WAR GAME)

2-23. When planning a breaching operation, war gaming is critical to refine the COA, including—

- Task-organizing support, breach, and assault forces.
- Developing a DST, identifying triggers to support obscuration, committing breach and assault forces, and shifting fires.
- Identifying and/or confirming the locations of decision points, NAIs, targeted areas of interest (TAIs), and other information that is needed to support decision points.
- Refining CCIR and incorporating them into the ISR plan/graphics.
- Refining the fire-support plan/graphics, including the obscuration plan.
- Confirming the obstacle template according to the TF R&S plan.

2-24. The commitment of the breach force is a critical decision point that must be included in the DST. The commander must thoroughly analyze what criteria must be met before the breach force begins obstacle reduction. Commitment criteria elements may include the following:

- Destruction of certain vehicles or a certain number of vehicles.
- Effective suppression of the enemy by the support force.
- Effective obscuration of the enemy.
- Remaining strength of the support force.
- Remaining reduction assets available to the breach force.
- Activation of CFZ.
- Air-defense assets in position.

NOTE: A plan must exist on how information will be gathered to verify the commitment criteria.

2-25. Once war gaming is complete, a COA has been refined, and a final task organization has been determined, a C^2 system can be planned to support the operation. To conduct a successful breaching operation, an effective C^2 system must be established. Specifically, each force in the breaching organization (support, breach, and assault) must understand who is in charge, especially within support and breach forces.

2-26. The support force may be comprised of more than one element. For example, in a TF breaching operation, the support force may be comprised of two company teams. Due to the importance of the support force's role and its size, it may be best to have the TF commander/S3 be the support force commander. He is responsible for coordinating the actions of both company teams and determining when the conditions have been met for the breach force's commitment. This allows each company commander to concentrate on directing his unit while under the overall control of an experienced leader.

2-27. The breach force, with its two subordinate elements (security and reduction), must have a well-understood C^2 system in place. There are many ways to establish C^2 of the breach force, and the method used may depend on the—

- Size and composition of security and reduction forces.
- Number of lanes that must be created.
- Experience level of the personnel within the breach force.

2-28. The following are possible C^2 structures within the breach force for a TF breaching operation:

- A company-team commander acts as the security element commander and the breach force commander, with an engineer as the reduction element commander.
- An engineer-company commander acts as the reduction element commander and the breach force commander, with a maneuver platoon leader as the security element commander.

2-29. Another C^2 issue that must be resolved during COA analysis is traffic control through lanes created in an obstacle system. *Figure 2-1* shows an example of the graphic control measures that may be established to facilitate the efficient movement of forces through an obstacle system. In this example, traffic-control post (TCP) 8 is located at the far-recognition marker. These control measures can be initially established by the higher HQ but need to be refined as the unit conducting the breaching operation develops its scheme of maneuver. *Table 2-1* lists the responsibilities for creating lanes and controlling traffic through an obstacle system when a TF breaching operation is being conducted.

PREPARATION

2-30. Intelligence collection continues throughout preparations for a breaching operation. As intelligence reports are received, the SITEMP and the ISR plan are updated and revised. Intelligence reports are also used to refine the task organization of support, breach, and assault forces and the scheme of maneuver, including proposed points of breach and target grids for smoke and suppressive fires. Updated intelligence information is also used during combined-arms rehearsals.

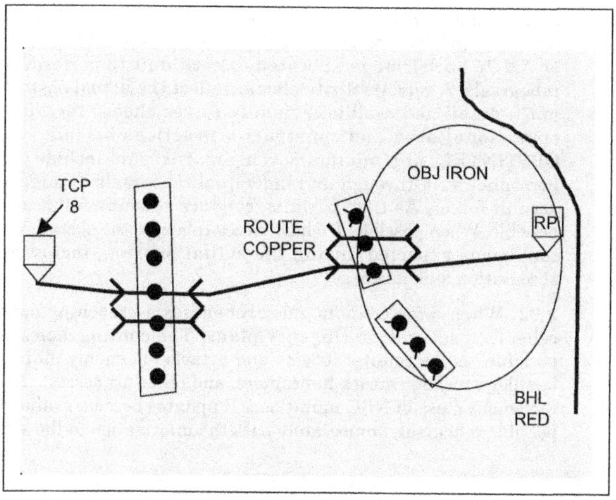

Figure 2-1. Graphics to Support Passage of Follow-On Forces

Table 2-1. C^2 Responsibilities for Lanes and Traffic Control

Breaching Operation	Responsibilities of TF Forces	Responsibilities of the Brigade
Initiating breaching operations through the passage of the assault force through tactical obstacles	The breach force creates lanes for the assault force (reduces, proofs, and emplaces the initial lane-marking pattern and assists in the passage of the assault force through the marked lanes). The TF controls all traffic passing though tactical obstacles.	
Completing the passage of the assault force through the seizure of the farside objective	The assault force creates and marks lanes through protective obstacles up to the BHL.	Assumes control of lane(s) in tactical obstacles Creates more lanes as necessary Upgrades lane marking to the intermediate lane-marking pattern Ensures that lanes remain open Assumes traffic control through tactical obstacles
Passing follow-on forces through tactical and protective obstacles to the BHL	The assault force assists in the passage of follow-on forces (from tactical obstacles to the BHL).	
Completing the forward passage of lines		Assumes control of all lanes and traffic up to the BHL

2-31. A unit meticulously plans, manages, and controls breaching rehearsals. (For more information on breaching rehearsals, see *Chapter 4* of this manual or *FM 101-5*.) Time is allocated to each unit to perform combined-arms rehearsals. A rehearsal site should reflect the actual obstacle system in as much detail as possible. Friendly forces choose terrain similar to the operational area and construct a practice obstacle system based on OBSTINTEL. At a minimum, rehearsals should include a leader and key personnel walk-through and individual rehearsals by support, breach, and assault forces. As time permits, conduct as many full-scale rehearsals as possible. When possible, friendly forces rehearse the operation under the same conditions expected during the actual mission, including battle-space obscuration and darkness.

2-32. When a force commander rehearses a breaching operation, he also rehearses several contingency plans. The contingencies should include possible enemy counterattacks and attacks by enemy indirect-fire systems (artillery, rockets, attack helicopters, and other air assets). This also includes the enemy's use of NBC munitions. If updates become available after the last possible rehearsal, immediately pass the information to the affected elements.

EXECUTION

2-33. Each element of the breaching organization must accomplish its assigned tasks for a breaching operation to be successful. Additionally, the entire force must be flexible when responding to changes in the plan. Examples of changes are—

- Allocating additional assets to the support force due to attrition.
- Changing the location of the point of penetration.
- Modifying the scheme of reduction and changing the order of the units passing through the created lane.

2-34. All units must continually update the commander during mission execution. A critical phase of a breaching operation is the movement of the support force from the LD to its SBF positions. The support force must be prepared to maneuver to its SBF positions. Additionally, the support force must plan for the enemy's use of SCATMINEs along its axis of advance, which may require it to conduct a breaching operation.

2-35. After the support force has occupied its SBF positions and the commitment criteria of the breach force has been met (achieved necessary suppression and obscuration), the breaching commander should order the breach force to move forward to a specific point (6- or 8-digit grid) on the battle space to begin reduction. As the breach-force commander maneuvers his force to the point of breach, he must always remain cognizant of his relationship to the support force to ensure that he does not unnecessarily mask their fires.

2-36. Actions by the assault force at an assault position may include—

- Verifying current friendly and enemy situations using tactical reports from company teams or support forces.
- Issuing fragmentary orders (FRAGOs) and disseminating information to the lowest level.

- Confirming TRPs and direct-fire responsibilities.
- Positioning artillery observers.
- Conducting final prepare-to-fire checks.
- Reorganizing to compensate for combat losses.

2-37. The breach force commander should lead with his security element to allow it to establish nearside security before the reduction element arrives at the obstacle. As the security element occupies its position, the reduction-element commander should conduct a recon to acquire obstacle information and quickly confirm/refine his scheme of reduction. As the reduction-element commander brings his element forward, he should report the exact location where the obstacle will be reduced and the estimated time to reduce/proof and mark the obstacle. At the same time, the breach-force commander must assess the effectiveness of the suppression and determine if and how he will augment the support force's fires.

2-38. As the reduction continues, the support- and breach-force commanders must update their commander with the current status. Specifically, the support-force commander must update on his ammunition status and the strength of his unit. The breach-force commander must update on the progress of the reduction effort and the effectiveness of obscuration and suppression. The breach-force commander should verify who the assault force will be and where the assault force will come from so that he can assist in their passage.

2-39. As the reduction effort nears completion, the breach-force commander should report the grid coordinates of the far-recognition marker so that the assault force can begin movement to that location. When reduction is complete, the breach-force commander must—

- Report lane completion.
- Provide grid coordinates of the final-approach marker.
- Confirm/report the lane-marking pattern and material.

2-40. The breach-force commander establishes farside local security and assists in the assault force's passage. Once the lanes have been reported/signaled as being open for traffic, the assault force passes through the lanes to destroy forces that can bring direct fires on created lanes. To accomplish this mission, the assault force may have to reduce enemy protective obstacles.

2-41. The obstacle system continues to be a choke point and danger area even after the assault force has passed through created lanes. Additional lanes are constructed to speed the follow-on forces' passage, and lanes are monitored to ensure that they remain clear. Reduction assets need to be maintained near the points of breach in case a scatterable minefield is emplaced by the obstacle. The lane-marking pattern is upgraded to intermediate. There are many ways this can be accomplished. The higher HQ—

- Takes command of the point of breach and has some or all of the reduction element revert to its control and continue creating/expanding lanes in the obstacle for follow-on forces.

- Takes command of the point of breach and has additional reduction assets follow closely behind the assault force and create additional lanes to allow the reduction element to remain under unit control.
- Tasks the breaching unit to maintain lanes in tactical obstacles.

2-42. The lanes are widened to allow two-lane traffic through obstacles, and they are marked with the full lane-marking pattern. Deliberate marking and fencing systems are installed, and military police (MP) establish necessary traffic control. Eventually, follow-on engineer forces clear obstacles and eliminate the choke point.

Chapter 3

Lane and Bypass Marking

This chapter implements STANAG 2036.

This chapter provides commanders with an Army-wide standard system for lane and bypass marking. It centers around a systematic, phased upgrade of lane marking. Each upgrade conforms to tactical requirements for that phase of the attack, from initial obstacle reduction to the passage of larger follow-on forces and the return traffic necessary to sustain the force. Lane or bypass marking is a critical component of obstacle reduction. Effective lane marking allows commanders to project forces through an obstacle quickly, with combat power and C^2 intact. It gives an assault force and follow-on forces confidence in the safety of the lane and helps prevent unnecessary casualties.

LANE-MARKING SYSTEM

3-1. There are two critical components to any lane-marking system—the lane-marking pattern and the marking device.

- **Lane-marking pattern**. The location of markers indicating the entrance, the lane, and the exit.
- **Marking device**. The type of hardware emplaced to mark the entrance, the lane, and the exit.

3-2. The lane-marking system outlined in this chapter centers on standardizing the marking pattern across the Army rather than standardizing the marking device. Standardizing the marking pattern is critical to offensive operations. A common lane-marking pattern—

- Enables cross attachments and adjacent units to recognize lanes easily with minimal knowledge of a particular unit's tactical SOP.
- Gives all forces a standardized set of visual cues needed to pass through a lane safely and maintain their momentum.
- Facilitates quick conversion to the lane-marking requirements of STANAG 2036.

3-3. Until the Army adopts a standard lane-marking device, commanders decide what hardware to use. This gives units greater flexibility, allowing them to adopt a marking device that is tailor-made for their type of unit and mission (armored or light force, mounted or dismounted attack, limited visibility, thermal capability). Regardless of the device used, it must support the standard lane-marking pattern outlined in the following paragraphs. Therefore, commanders should consider the marking guidelines and examples in this chapter before developing or adopting their own system. Commanders should determine and standardize the marking device used based on the availability of resources and METT-TC.

LANE-MARKING TERMS

3-4. The following terms provide a common basis for discussing lane marking:

- Entrance markers.
- Handrail markers.
- Exit markers.
- Entrance-funnel markers.
- Final-approach markers.
- Far-recognition markers.
- TCPs and guides.

ENTRANCE MARKERS

3-5. Entrance markers indicate the start of a lane through an obstacle. They signify the friendly-side boundary of the obstacle and the point at which movement is restricted by the lane width and path. Entrance markers also indicate the lane width. Therefore, it is critical for a maneuvering force to distinguish the entrance point clearly, since it signifies the point at which passing vehicles can no longer adjust their movement in reaction to the situation (direct and indirect fires) without jeopardizing the force. Entrance markers must be visually different from handrail markers to help a force distinguish this critical point in the lane. The distance between entrance markers must be the same as the width of the created lane. Entrance markers placed a minimum of 4.5 meters apart indicate a lane capable of supporting mounted movement, while markers placed a minimum of 1 meter apart indicate a dismounted lane. When obstacle boundaries are vague or unknown, such as a buried minefield, the breach force uses its best judgment and marks the entrance where obstacle reduction and lane proofing begin.

HANDRAIL MARKERS

3-6. Handrail markers define the lane path through an obstacle and indicate the limits of the lane width. As a minimum, mounted and dismounted lanes have a left handrail. Mounted and dismounted forces moving through a lane should keep the left handrail immediately to the left of their vehicle or person. The lane width is defined by the entrance markers. Therefore, when only the left handrail is marked, drivers use entrance and handrail markers to gauge the lane width and path. As the operation progresses, lane marking may be upgraded to include left and right handrails.

EXIT MARKERS

3-7. Exit markers indicate the far side limit of a lane through an obstacle. For a passing force, the exit marker signifies the point at which movement is no longer confined to the lane path. Like entrance markers, exit markers must be distinguishably different from handrail markers; however, the exit may be marked the same as the entrance. Exit markers are placed to the left and right of an exit point and spaced the width of the reduced lane. This visual reference is critical when only a left handrail is marked. The combination of entrance, left-handrail, and exit markers gives the driver and the tank commander visual cues (entrance and exit points, lane width and path) to pass safely along a lane.

ENTRANCE-FUNNEL MARKERS

3-8. Entrance-funnel markers augment entrance marking. The V formed by the funnel markers forces approaching platoons into a column formation and assists drivers and tank commanders in making last-minute adjustments before entering the lane.

FINAL-APPROACH MARKERS

3-9. Final-approach markers are highly visible, more robust markers that augment the visual signature of entrance-funnel markers. Units must be able to see the entrance funnel and/or the entrance-funnel markers from the final-approach marker. The final-approach marker—

- Provides the assault-force commander with a highly visible reference point toward which to maneuver his formation.
- Signals company-team commanders to begin changing to a column formation.

FAR-RECOGNITION MARKERS

3-10. Far-recognition markers are highly visible markers located between the final-approach marker and the friendly unit. They are primarily used when passing battalion-size forces through a lane where direct observation of the final-approach marker is denied due to distance, visibility, or terrain. Far-recognition markers should be different from the final-approach marker. Far-recognition markers indicate the point at which forces begin changing their formation to posture for the passage. A single far-recognition marker may serve up to two lanes when located 200 to 400 meters apart. Once lanes are upgraded to two-way traffic, far-recognition markers are required for each two-way lane. Far-recognition markers should be visually alterable so that they can be distinguished from the far-recognition marker of an adjacent lane. This assists the C^2 of large formations when passing on several adjacent lanes. When a far-recognition marker serves more than one lane, a guide or a TCP is collocated with the far-recognition marker nearest the obstacle.

TRAFFIC-CONTROL POSTS AND GUIDES

3-11. A TCP or a guide is a two-man team with communications means that assists the commander in controlling the movement of forces. When available, consider using MPs to establish and man these posts. Initially, the commander may use other personnel as guides to man critical far-recognition markers until MPs establish full TCPs. TCPs and guides provide the commander with a means to control traffic flow to appropriate lanes. When there are multiple lanes branching off a single far-recognition marker, the TCP helps direct the formation to various lanes. The TCP can also assist in modifying the traffic flow when lanes have been closed for maintenance, for lane expansion, or by enemy SCATMINEs. The TCP or guide must give the commander the azimuth and distance to the final-approach marker, the device used for the approach marker, and the level of the lane-marking pattern. For light forces, guides may physically escort passing units from the far-recognition marker to the lane entrance. In short, the TCP allows the commander to make last-minute changes in the traffic flow, giving him increased flexibility to react to the enemy situation.

3-12. Another technique to facilitate the movement of the assault force to the points of breach is for the assault force to maneuver a recon element, such as a section, behind the breach force. This section observes the breach effort, identifies lanes, and leads the assault force to lanes after the breach is complete.

LEVELS OF LANE MARKING AND PATTERNS

3-13. There are three levels of lane marking—initial, intermediate, and full. Each lane-marking level provides an increase in lane signature and capability. Lane requirements change as a breaching operation matures from the passage of the assault force to the passage of larger follow-on forces. Initial lane-marking requirements are driven by the nature of the fight through the obstacle. Marking must be rapid, providing only the bare minimum signature needed to safely pass small units (company teams, platoons) that make up the assault force. This contrasts with lane requirements during later phases of an attack where larger units (battalion and above) are passed to subsequent objectives and mark improvements to be made. Two-way traffic becomes a priority for the simultaneous forward passage of combat units and return traffic (ambulances, empty supply vehicles) necessary to sustain the force. With the increase in traffic volume comes more diverse forces and levels of driver experience. Lane-marking limits must be clear to the most inexperienced driver or crewman; do not assume he has a knowledge of the unit's SOP. A fully developed lane must support two-way traffic and be completely marked.

3-14. Commanders must be aware of how the lane needs of a force change with the operation so that they can anticipate lane-marking and lane-capability requirements. Integrating the levels of lane marking into the overall breaching plan ensures that a unit's needs are satisfied. The forces necessary to mark and upgrade lanes must be allocated and tasked with that mission. The scheme of maneuver and the service-support plan are the basis for analyzing lane requirements. The following paragraphs describe lane-marking patterns in detail and provide guidelines on when a commander should upgrade lane marking and lane capability.

INITIAL LANE MARKING

3-15. The breach force emplaces the initial lane-marking pattern immediately after the obstacle is reduced and, if required, the lane is proofed. This is a signal to the assault-force commander that the lane is ready for traffic. Initial lane marking is kept to a minimum, focusing on the markings needed to pass immediate assault forces (company teams and smaller) through the lane to seize the point of penetration on the farside objective. Normally, the assault force can observe the lane as it is being marked and does not need the greater visual signature of mature lane marking. The initial lane-marking pattern contains—

- Entrance markers.
- Left-handrail markers.
- Exit markers.
- Entrance-funnel markers.
- A final-approach marker.

3-16. *Figure 3-1*, page 3-6, shows the initial lane-marking pattern. Entrance, left-handrail, and exit markers are the first markers that the breach force emplaces since these markers define the location and limits of the lane. These markers should be emplaced as follows:

- Entrance markers are placed to the left and right of the reduced lane's entrance point. They are spaced the width of the lane (at least 4.5 meters for mounted lanes and 1 meter for dismounted lanes).
- Left-handrail markers are placed at the left limit of the lane along the entire path. They are placed at 15-meter intervals for mounted forces and 5-meter intervals for dismounted forces. Commanders can modify the intervals based on the terrain, the visibility, the lane length, and the lane path. Additionally, commanders may choose to mark left and right handrails when the lane path is lengthy or unclear (through a complex obstacle).
- Exit markers are placed to the left and right of the lane at the far side of the obstacle (at least 4.5 meters for mounted lanes and 1 meter for dismounted lanes).

3-17. After entrance, left-handrail, and exit markers are placed, the breach force emplaces entrance-funnel markers and the final-approach marker. These markers enhance the visual signature of the lane entrance for the assault force. If the assault force is in a position to observe the obstacle reduction, the commander does not have to wait until entrance-funnel markers and the final-approach marker are set to commit the assault force. However, the breach force continues to install the necessary markers to improve the lane signature for larger or subsequent follow-on forces maneuvering toward the lane. Entrance-funnel markers and the final-approach marker are emplaced as follows:

- Entrance-funnel markers are placed at 15-meter intervals for mounted forces and 5-meter intervals for dismounted forces. They are placed diagonal to the lane entrance to form a 45-degree V as shown in *Figure 3-1*.
- The final-approach marker is centered on the lane and placed at least 200 meters from the lane entrance for mounted forces. For dismounted forces, the nature of the attack may preclude using a final-approach marker initially. Place a final-approach marker 30 meters from the entrance as soon as the mission allows. Commanders modify the recommended distance for the final-approach marker based on the terrain and the visibility.

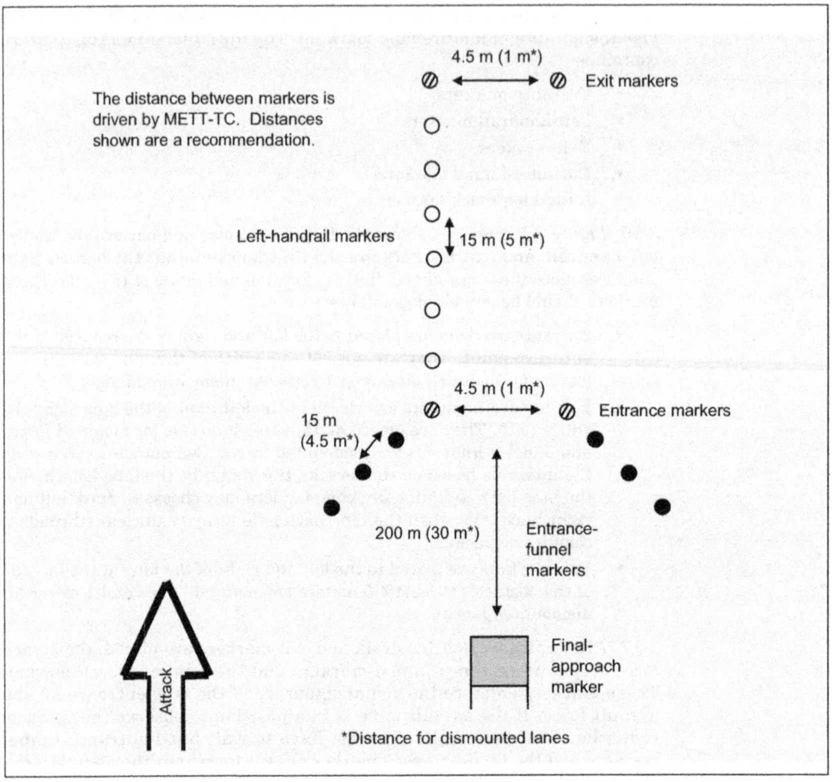

Figure 3-1. Initial Lane-Marking Pattern

INTERMEDIATE LANE MARKING

3-18. Upgrading initial lane marking to intermediate lane marking is triggered by one of two key events—

- Commitment of larger combat forces who are unable to directly observe the obstacle as it is reduced.
- Rearward passage of sustainment traffic (casualty evacuation and vehicle recovery).

3-19. Intermediate lane marking has two goals—

- Increasing the lane signature to assist in the passage of larger combat forces.
- Providing sufficient marking for two-way, single-lane traffic.

3-20. Intermediate lane marking builds on the initial lane-marking pattern by adding right-handrail markers, exit-funnel markers, far-recognition markers, and a farside final-approach marker. *Figure 3-2* shows the intermediate lane-marking pattern.

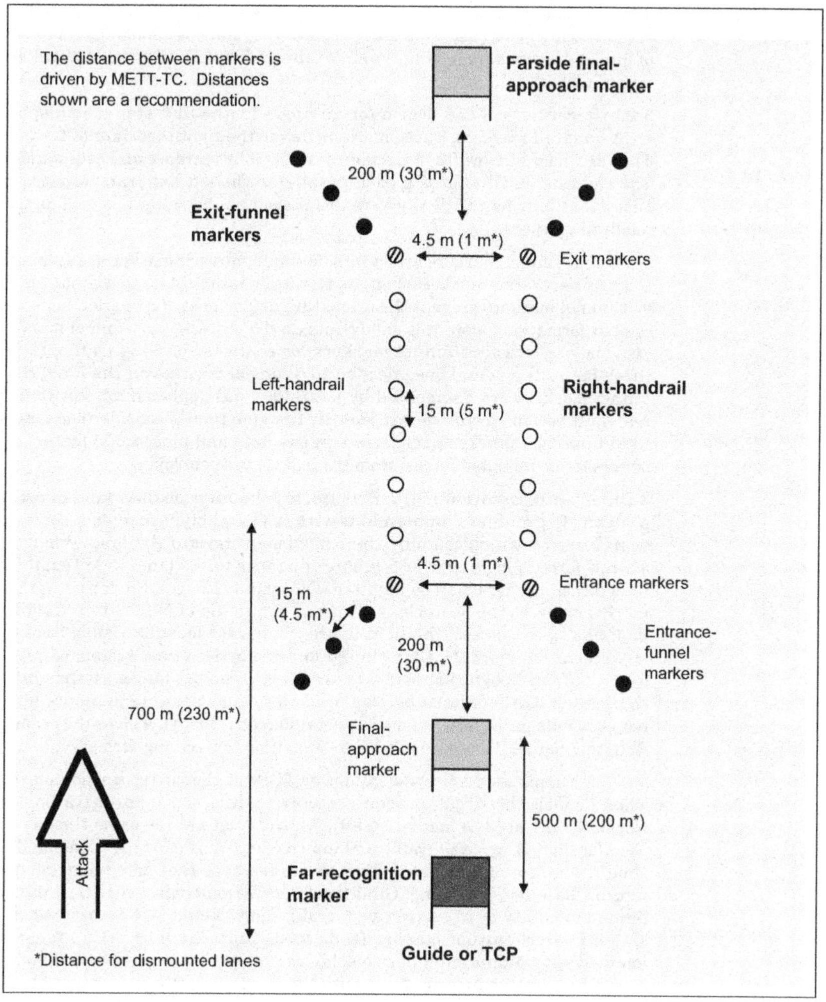

Figure 3-2. Intermediate Lane-Marking Pattern

3-21. The commander sets the priority of marker emplacement based on the situation. If the scheme of maneuver requires the immediate passage of larger combat forces, right-handrail markers and the far-recognition marker may be the priority. On the other hand, if it is necessary to ground-evacuate casualties or to recover vehicles, it may require emplacing right-handrail markers, exit-funnel markers, and the farside final-approach marker first. Intermediate lane marking gives the commander greater flexibility. Establishing a network of guides becomes increasingly important to direct the flow of forward and returning forces.

3-22. Regardless of the reason for an upgrade, the first step is to emplace right-handrail markers. These markers define the rightmost limit of the lane. They are placed the width of the lane as defined by entrance and exit markers. The right handrail follows a path parallel to the left handrail through the obstacle. Right-handrail markers are placed at the same interval as left-handrail markers.

3-23. Exit-funnel markers and a farside final-approach marker are emplaced to mirror entrance markers. For the forward passage of large combat forces, exit-funnel markers prevent the premature deployment of a passing force into combat formation before it is safely outside the obstacle. Exit-funnel markers also become entrance-funnel markers for rearward passing traffic, giving these forces the visual cues needed to align themselves on the lane. Exit-funnel markers are augmented by a farside final-approach marker to help rearward passing forces clearly identify the lane from their side. The farside final-approach marker is centered on the lane and placed 200 meters (30 meters for dismounted forces) from the exit (farside entrance).

3-24. A far-recognition marker completes the intermediate lane-marking pattern. It provides commanders with a visual signature or a series of signatures on which to guide their movement toward the lane. When the assault force is moving over a greater distance to the lanes, additional far-recognition markers may be required. For mounted forces, the far-recognition marker nearest the obstacle is placed a maximum of 500 meters from the final-approach marker (the final-approach marker must be visible from the far-recognition marker). Dismounted forces may require a system of guides instead of far-recognition markers for passing combat forces. However, far-recognition markers must be emplaced as soon as possible to reduce guide requirements for passing mounted sustainment traffic. This gives the assault-force commander the space needed to transition to a column formation.

3-25. A commander collocates guides or TCPs at the far-recognition marker when he feels the situation requires more positive control over traffic flow. Normally, the need to increase traffic control comes at the same time as the need for limited two-way traffic and intermediate lane marking. Commanders should plan for the use of dedicated guides once they have upgraded to intermediate-level marking. Guides or TCPs become mission critical during limited visibility or in restrictive terrain. They should also be used when a single far-recognition marker feeds more than one lane. TCPs must be manned with a minimum of two soldiers and must have secure frequency-modulated (FM) communications with the controlling HQ. It is essential that soldiers acting as guides or TCPs know the—

- Azimuth and distance to the final-approach marker or the 8-digit grid coordinate of the final-approach marker that is entered into the unit's Global Positioning System (GPS) receiver.
- Lane-marking pattern currently emplaced.
- Type of final-approach marker used.
- Traffic-control plan and march order.

Guides must be kept up-to-date on the status of lane marking, maintenance, and closures.

FULL (TWO-WAY) LANE MARKING

3-26. Lane marking is upgraded to a full lane-marking pattern when the commander desires uninterrupted two-way traffic flow. Expanding lanes to full lane marking is resource-intensive and is not part of a breaching operation. A full lane supports uninterrupted two-way traffic. The marking pattern is modified to give forces passing forward or rearward the same visual signature.

3-27. Upgrading to a full lane-marking pattern is normally assigned to follow-on engineer forces, since it is usually beyond the immediate capability of engineers with forward units. In special cases, the commander may task engineers with forward units to expand and upgrade a lane from an intermediate to a full lane-marking pattern. However, the trade-off is decreased support to forward units and delays associated with changing task organization. The full lane-marking pattern is also used when marking a lane through friendly obstacles along a main supply route (MSR) or a passage lane.

3-28. Upgrading intermediate lane marking to full lane marking begins by temporarily closing the lane, rerouting traffic, and expanding the lane width. The lane is always expanded to the left in relation to the direction of the attack. Engineers reduce and proof the obstacle beginning at the left handrail to give a total lane width of 10 meters (5 meters each way). The expansion width requirement is the same for armored and light forces, since both forces must be able to pass mounted sustainment and combat forces. When the lane is expanded by mechanical means, it is necessary to entirely remove the left-handrail markers during expansion. Whenever possible, these markers should be left alone to serve as a reference point during lane expansion. If removed altogether, left-handrail markers must be replaced since they form the left handrail for both forward and rearward passing forces.

3-29. When the lane width has been expanded to 10 meters, entrance, exit, right-handrail, funnel, and final-approach markers are replaced on the return lane. All markings are the same as described in previous paragraphs. *Figure 3-3*, page 3-10, shows a full lane-marking pattern.

3-30. The full lane-marking pattern has three entrance and three exit markers. Entrance and exit markers are placed the width of forward and return lanes and are visually different from other markers. Three entrance markers indicate a two-way traffic lane, and units should always use the rightmost lane. Entrance- and exit-funnel markers are placed slightly different from previous marking patterns. In the full lane-marking pattern, funnel markers extend out from the entrance and exit markers on the outside

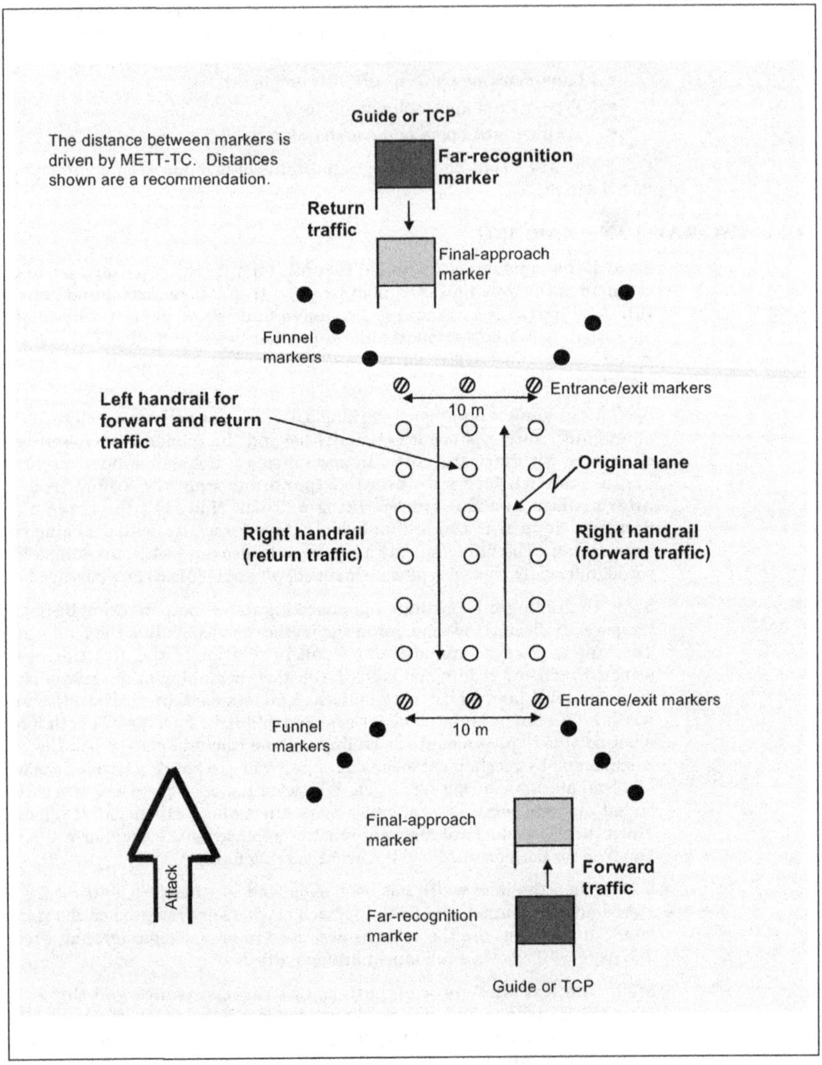

Figure 3-3. Full Lane-Marking Pattern

only. Final-approach markers are placed 200 meters from and centered on entrances of forward and return lanes. This helps forces identify entrance points from either direction. Far-recognition markers are placed a maximum of 500 meters from forward and return final-approach markers. Again, distances may be modified based on the terrain, the visibility, and the mission.

COMMANDER'S GUIDANCE FOR LANE MARKING

3-31. *Table 3-1* provides the commander with a summary of lane-marking levels, guidelines on unit responsibilities, and events that trigger a lane-marking upgrade. The chart focuses on TF-level breaching operations. In the table, *who* refers to the unit responsible for lane marking, *when* describes events that trigger the need to upgrade, and *markers* highlight a quick reference for lane-marking requirements.

Table 3-1. Lane-Marking Levels, Unit Responsibilities, and Trigger Events

	Initial	Intermediate	Full (Two Way)
Who	TF breach force	TF breach force	Brigade
When	Obstacle is reduced	Passing battalion- or company-size forces	Passing brigade- or battalion-size forces
	Passing platoon- or company-size assault forces	Passing force which cannot see the lane	Situation requires uninterrupted sustainment traffic
		Passing TF combat trains	
Markers	Entrance	Add right handrail	Expand lane width to 10 meters
	Exit	Add exit funnel	Adjust entrance/exit
	Left handrail	Add farside final approach	Adjust left/right handrails to new width
	Entrance funnel	Add far recognition	Add far recognition
	Final approach	Add guides or TCPs	Add farside guides or TCPs

MOVEMENT THROUGH THE LANE

3-32. Standard lane-marking patterns help the commander in two critical aspects when moving a unit through a lane:

- Lane markers help the force posture its formation for efficient passage through the defile that is caused by the lane through the obstacle.
- The combination of lane markers and TCPs gives the commander greater flexibility in the forward and rearward movement of forces.

3-33. Far-recognition and final-approach markers in a standard lane-marking pattern help commanders smoothly change assault and follow-on forces into a column formation before passing through the lane. In most cases, the situation should allow the unit to move in a column formation from the outset; however, the enemy situation and the terrain may demand that forces remain dispersed as long as possible. Far-recognition, final-approach, and entrance-funnel markers are used to trigger commanders to alter their formations.

Figure 3-4 shows how formations at various levels may change to efficiently pass through lanes. A battalion TF initially moves from its assault position in a box, V, or wedge formation depending on the number of lanes being used.

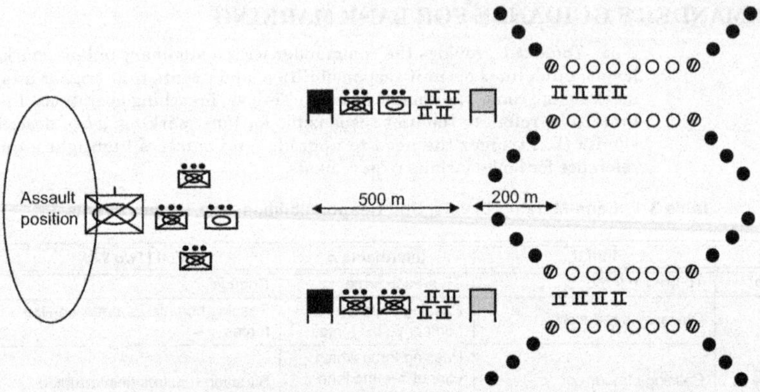

	Move out of assault position	Visually identify far-recognition marker	Pass far-recognition marker	Pass final-approach marker	Pass through exit-funnel markers
Battalion	Box (2 lanes) V (3 lanes) Wedge (3 lanes)	Box (2 lanes) V (3 lanes) Wedge (3 lanes)			
Company	Any combat formation	Column with dispersal	Column	Column	Resume formation
Platoon	Any combat formation	Any combat formation	Column with dispersal	Column	

Figure 3-4. Formation Changes Based on Lane Markers

3-34. Initially, company teams and below may move in any combat formation. Once lead elements of the battalion visually identify the far-recognition marker, company teams begin moving into a column formation with platoons still deployed. Platoons—

- Change into a column formation with dispersal as they move from the far-recognition marker to the final-approach marker.
- Change into a column formation as they move from the final-approach marker to entrance markers, using the entrance-funnel markers as guides.
- Redeploy once they pass through the exit or the exit funnels.

3-35. As the operation progresses, commanders should upgrade lane marking and establish traffic control as early as METT-TC allows. The combination of upgrading lanes from the initial marking pattern and establishing TCPs gives the commander greater flexibility. Normally, TCPs are established at forward and return far-recognition markers. Upgrading to full lane marking that is capable of supporting uninterrupted two-way traffic is generally determined by the need to pass increased volumes of traffic.

3-36. *Figure 3-5*, page 3-14, shows the flexibility that full lane marking and TCPs provide. In this example, friendly forces have created two lanes, 200 meters apart. Lane 1 is a full lane, and lane 2 is an intermediate lane. A far-recognition marker on the nearside and the farside can feed traffic to both lanes. Full lane marking and TCPs allow the commander to continue pushing combat power forward on lanes 1 and 2 simultaneously. At the same time, the farside TCP directs all return traffic to lane 1. Commanders should ensure that recovery vehicles are on site to remove immobile and destroyed vehicles.

LANE MAINTENANCE AND RESPONSIBILITIES OF TRAFFIC THROUGH LANES

3-37. As discussed in *Chapter 2*, units must plan for lane maintenance (widen lanes, upgrade lane markings, ensure that lanes remain open, ensure C^2 of traffic through the breach area). The responsibility for lane maintenance should be delineated in unit orders to ensure alignment with the scheme of maneuver. Specifically, units must determine who is responsible for—

- Maintaining lanes in tactical and protective obstacles.
- Maintaining lanes when a change occurs.
- Controlling traffic within the breach area.

3-38. *Figures 3-6 through 3-9*, pages 3-15 and 3-16, show how a unit defines responsibilities for lane maintenance and traffic control. In this example, a TF breaching operation is being conducted. Teams A and B provide suppressive fires as the support force, Team C reduces tactical obstacles as the breach force, and Team D seizes OBJs LION and TIGER as the assault force. The brigade intends to pass an armored TF through obstacles as the first element of the follow-on forces once the TF has cleared a route through the breach area. Note that the initial-lane requirement to pass Team D (the assault force) is one vehicular lane, but the breach force breaches a second lane to pass follow-on forces (an armored TF).

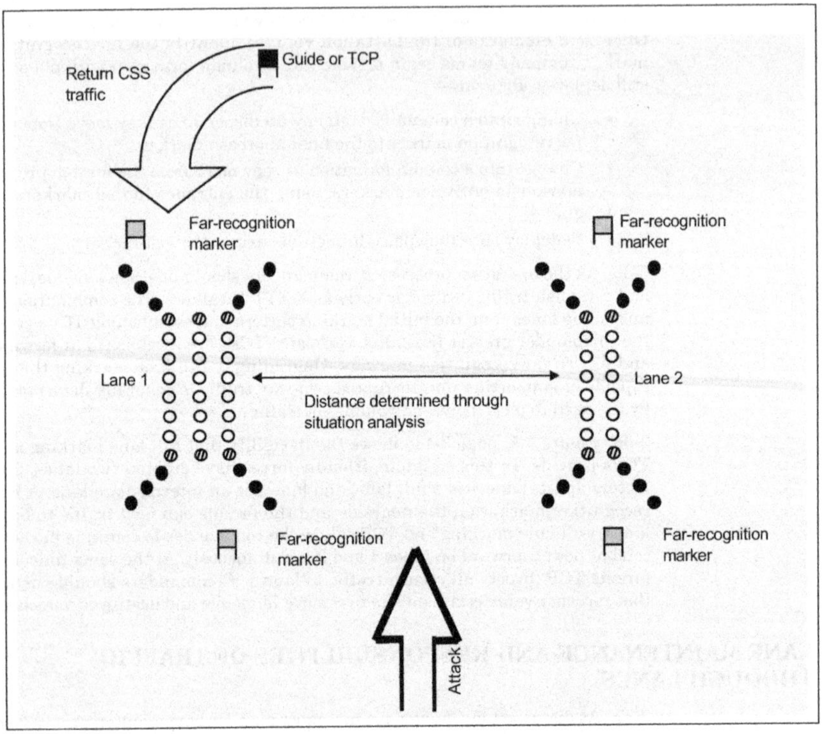

Figure 3-5. Multiple Lanes Used at Different Levels

3-39. As the breaching operation progresses, the breach force creates two lanes in the tactical obstacles. The breach force marks lanes with the initial lane-marking pattern and passes the assault force through the lanes (*Figure 3-6*). At this stage of the breaching operation, the TF maintains lanes in the tactical obstacles and controls all movement of forces within the breach area.

3-40. The assault force passes through the tactical obstacles and attacks to seize OBJs LION and TIGER. As the assault force attacks, it encounters protective obstacles and reduces and marks them. Immediately after the assault force passes through the tactical obstacles, the brigade assumes responsibility for lane maintenance in the tactical obstacles. The brigade also assumes responsibility of traffic control from the beginning of the breach area through the tactical obstacles. This includes creating additional lanes, ensuring that the lanes remain open, upgrading lane markings at the tactical obstacles, establishing TCPs, and calling units forward into the breach area. The TF is responsible for maintaining lanes through the protective obstacles (*Figure 3-7*).

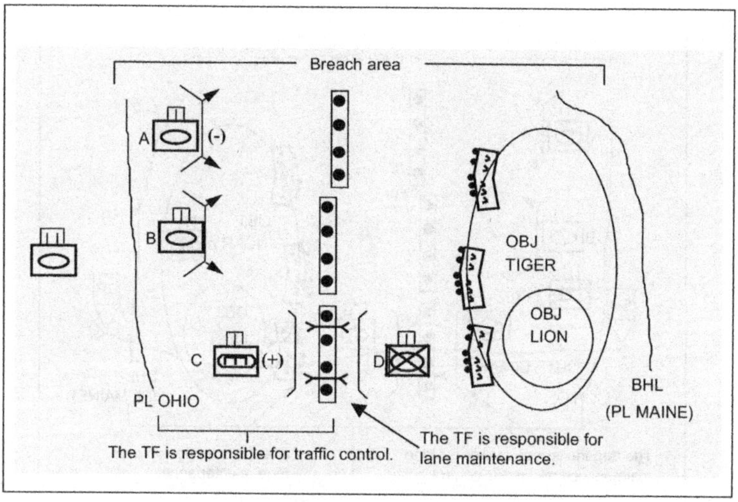

Figure 3-6. Assault Force Passes Through Tactical Obstacles

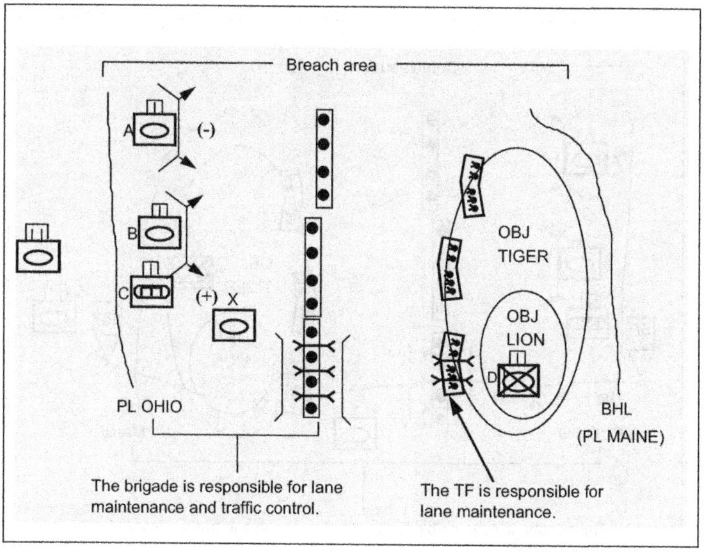

Figure 3-7. Brigade Assumes Control of Lanes in Tactical Obstacles

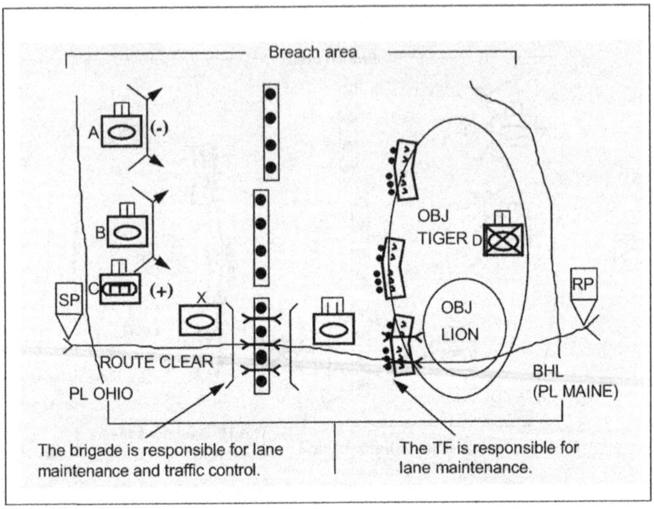

Figure 3-8. Follow-On Forces Begin to Pass Through the Breach Area

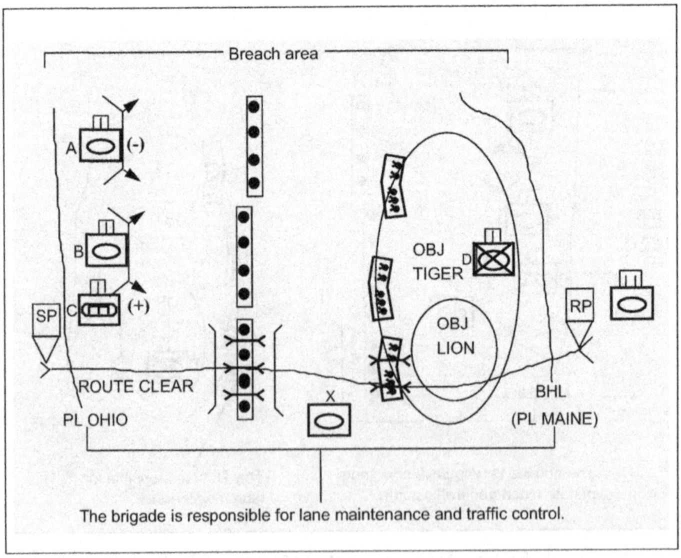

Figure 3-9. Brigade Assumes Full Control of the Breach Area

3-41. Once the assault force has seized its objective and reported that it has established a clear route up to the BHL (PL MAINE), the brigade begins to pass the follow-on TF through the breach area. At this point, the brigade controls traffic throughout the breach area. The TF is responsible for maintaining lanes in the protective obstacles (*Figure 3-8*).

3-42. As the follow-on forces cross the BHL, the breaching operation is complete; however, the obstacles within the breach area remain a constriction for other follow-on forces. At this point, the brigade maintains lanes and controls traffic. The brigade begins to focus on establishing MSRs through the breach area using corps engineer assets, which are task-organized to the brigade from the division (*Figure 3-9*).

LANE-MARKING DEVICES

3-43. Currently, the Army has no standard lane-marking device. The current marking systems in the Army's supply system are—

- Hand-Emplaced Minefield Marking System (HEMMS), national stock number (NSN) 9905-01-019-0140.
- Minefield Marking Set Number 2, NSN 9905-00-375-9180.
- Cleared-Lane Marking System (CLAMS), NSN 2590-01-205-3082.

3-44. Most lane marking in the field is done with nonstandard marking devices. Before adopting a nonstandard marking device, commanders should consider the guidelines summarized in *Table 3-2*. Coordination must occur between forces who are tasked to mark lanes and forces who are expected to use lanes to ensure that the marking material will work. For example, if a light force plans to create and mark vehicle lanes with an infrared light, vehicle drivers with enhanced (thermal) night-vision devices may not be able to see the markings.

Table 3-2. Guidelines for Lane-Marking Devices

Markers	Mounted Forces	Dismounted Forces
Handrail and funnel	Visible by TC and driver (buttoned up) from 50 meters	Visible by dismounted soldier in prone position from 15 meters
	Quick and easy to emplace (minimizes need to expose soldiers outside the emplacement vehicle)	Lightweight, quick and easy to emplace (dismounted soldier should be able to carry enough markers for the lane and still be able to fire and maneuver)
Entrance and exit	Visible by TC (buttoned up) from 100 meters	Visible by dismounted soldier from 50 meters
	Visually different from handrail and funnel markers	Visually different from handrail and funnel markers
	Quick and easy to emplace (may require soldier to dismount to emplace)	Lightweight, quick and easy to emplace
	Easily man-portable	
Final-approach and far-recognition	Visible by TC (not buttoned up) from 500 meters	Visible by dismounted soldier (on march) from 100 meters
	Visually different from each other	Visually different from each other
	Visually alterable to facilitate traffic control through multiple lanes	Visually alterable to facilitate traffic control through multiple lanes

3-45. *Figure 3-10* shows some of the devices that units currently use; they are easily procured or fabricated. This is by no means a complete listing; rather, it is intended to show commanders some of the options. Some of the general requirements for lane marking are—

- Markers must be able to withstand the rigors of the terrain, the weather, and the battle space.
- Markers should be easy to modify when visibility is limited, with minimal addition of manpower and equipment.
- Enhancements for limited visibility should be a constant source rather than a pulsating strobe. Strobes do not make the marking pattern readily apparent, particularly when approaching from an angle.

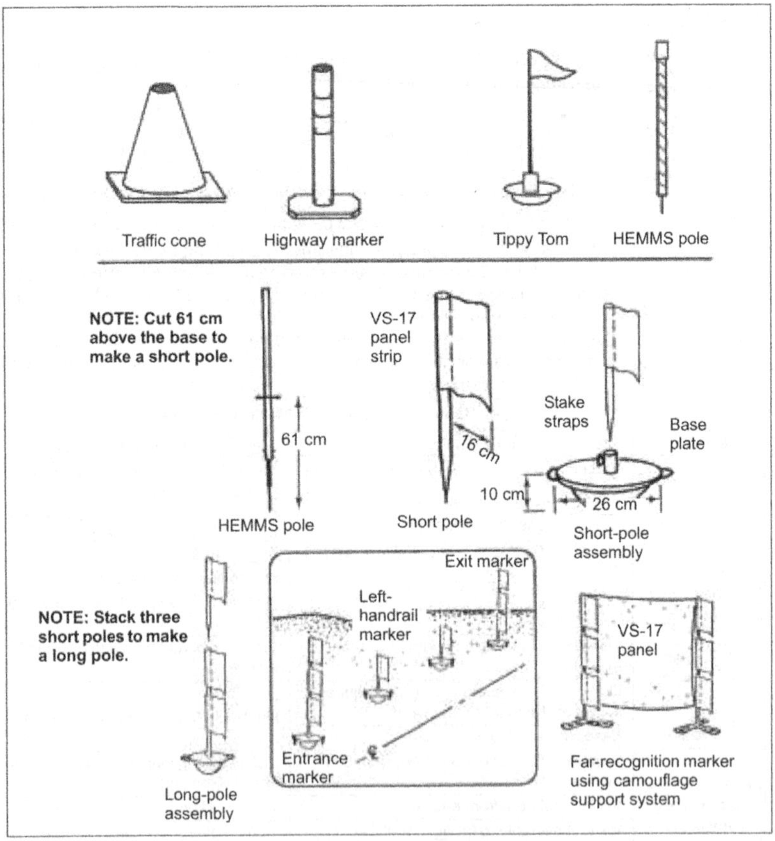

Figure 3-10. Examples of Nonstandard Marking Devices

BYPASS MARKING

3-46. Bypasses are marked using the same lane-marking patterns, marking devices, and visual cues as a lane. However, marking a bypass around an obstacle is much different from marking a lane through an obstacle because the amount of room available for forces to maneuver must be determined. While lanes through an obstacle are normally expanded to the left, bypasses are expanded away from the obstacle.

3-47. Bypasses must show where it is safe to rapidly pass around an obstacle, without unnecessarily restricting the force's ability to maneuver (*Figure 3-11,* page 3-20). To accomplish this, the marking unit will mark the bypass with a bypass-marking pattern (modified initial lane-marking pattern with right-handrail markers added). The unit will recon and initially mark the bypass 500 meters wide to allow platoons to maneuver around the obstacle. The width of the bypass is increased as time allows further recon. Limited visibility or additional obstacles discovered at the flank of the obstacle that is being bypassed may constrain the marked width of the path to less than 500 meters. (The right-handrail markers must be visible from the left-handrail markers.) In addition to bypass marking, units should consider establishing mobile guards on the friendly side of the obstacle to prevent vehicles/ personnel from entering it. All units on the battle space should possess marking material and understand how to emplace the bypass-marking pattern.

MARKING REQUIREMENTS UNDER THE NORTH ATLANTIC TREATY ORGANIZATION

3-48. The following paragraphs paraphrase the breach lane-marking requirements outlined in North Atlantic Treaty Organization (NATO) STANAG 2036. They also establish procedures used by US forces to modify intermediate and full lane-marking patterns to STANAG standard. The trend in modern warfare is increasing toward combined operations, so commanders need to be aware of their responsibilities under the NATO standard for marking hazardous areas, particularly lanes through obstacles.

COMMANDERS' RESPONSIBILITIES

3-49. STANAG 2036 outlines the minimum requirements for the lane-marking pattern before troops of other nations use the lane. Commanders must plan to convert the lane to the NATO marking standard as early as possible. When converting to the NATO standard, STANAG 2036 directs commanders to use the lane-marking devices as stated below. Within an offensive operation, this does not normally occur until the lane is matured to a full lane.

MARKING PATTERNS AND DEVICES

3-50. The intermediate lane-marking pattern satisfies the minimum lane-marking pattern that must be used before forces from another country pass through a lane. STANAG 2036 states that the entrance point, the exit point, and left and right handrails are the minimum required lane signatures.

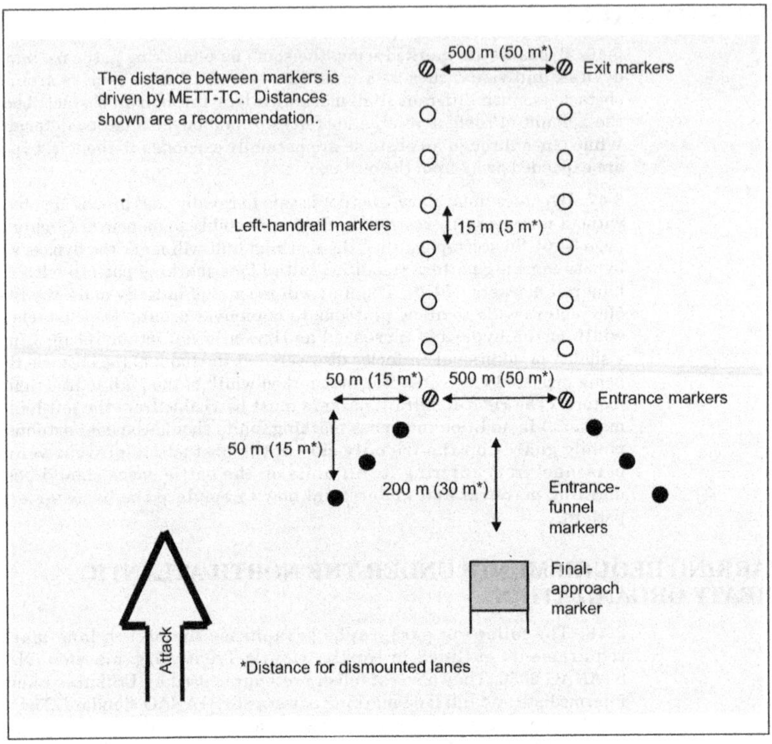

Figure 3-11. Bypass-Marking Pattern

Furthermore, STANAG 2036 requires that the entrance and exit of a lane be distinguishably different from other markers and that handrail markers be placed at intervals no greater than 30 meters. These requirements are met once lane marking is upgraded to the intermediate or full lane-marking pattern. Once the lane is marked to the the intermediate level, commanders can use the lane for allied forces to pass through without any additional marking if proper marking material is used.

3-51. *Figure 3-12* shows a NATO standard marker. The marker is placed at a right angle to the direction of travel so that the white portion of the arrow points inward to the lane, indicating the safe side on which to pass. The red portion points outward, indicating the lane limit or dangerous side of the lane. Units may fabricate NATO markers if they cannot get them through the supply system. STANAG 2036 requires that markers be visible from 50 meters under all conditions.

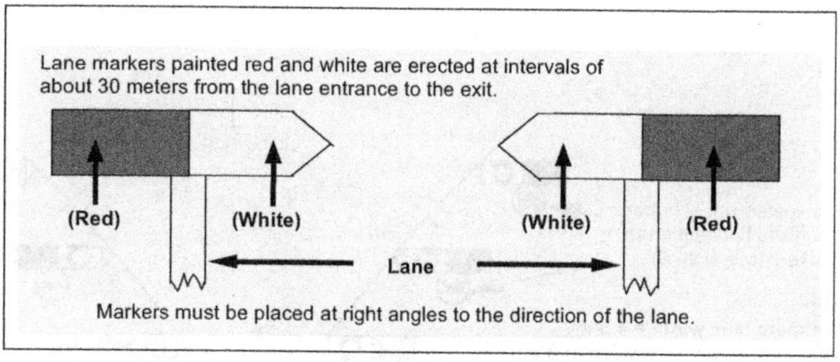

Figure 3-12. NATO Standard Marker

3-52. To convert intermediate lane marking to the NATO standard, affix NATO markers to long pickets and replace the existing entrance, exit, funnel, and handrail markers one for one. *Figure 3-13*, page 3-22, shows how NATO markers are used to convert an existing intermediate lane-marking pattern. Two NATO markers are used for entrance and exit markers to make them distinctly different. One NATO marker is affixed to each funnel marker and to each left- and right-handrail marker.

3-53. NATO uses white or green lights to illuminate markers at night. Entrance and exit markers are marked with two green lights or two white lights, placed horizontally, so that safe and dangerous markings on them are clearly visible. One white light or one green light is used on funnel and handrail markers. Commanders decide whether the light is placed on top of the NATO marker or placed so that it illuminates the marker (*Figure 3-14*, page 3-23). Lights must be visible from 50 meters under all conditions. Light sources for nighttime markers must have a continuous life of 12 hours.

COMPLEX OBSTACLES

3-54. When marking lanes through a complex obstacle, units can mark each obstacle independently or mark the entire obstacle system at once. The depth of the obstacle system, the distance between the obstacles, and the amount of marking material available determine which method is employed. See *Figures 3-15* and *3-16*, page 3-23, for examples of how complex obstacles may be marked; the examples use the initial lane-marking pattern.

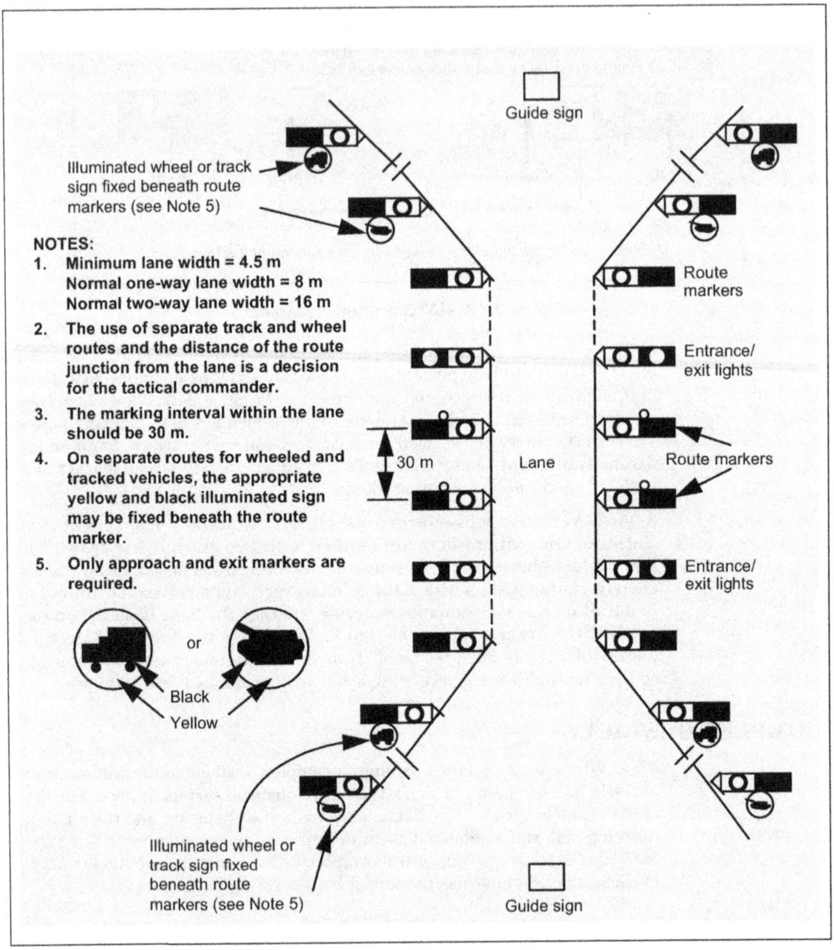

Figure 3-13. NATO Lane-Marking Standard

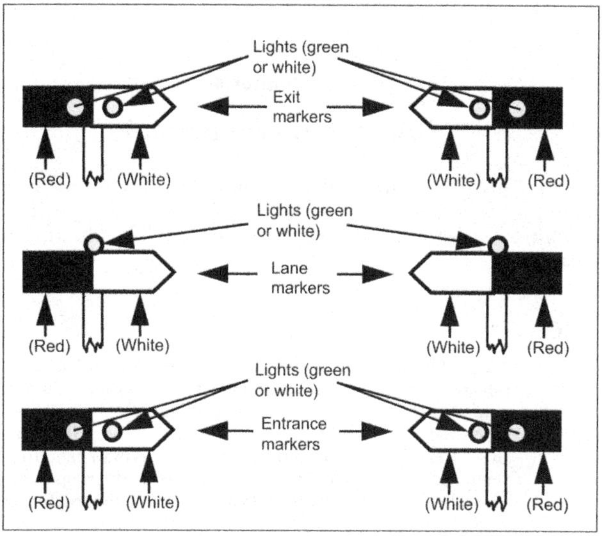

Figure 3-14. NATO Standard for Marking Lanes at Night

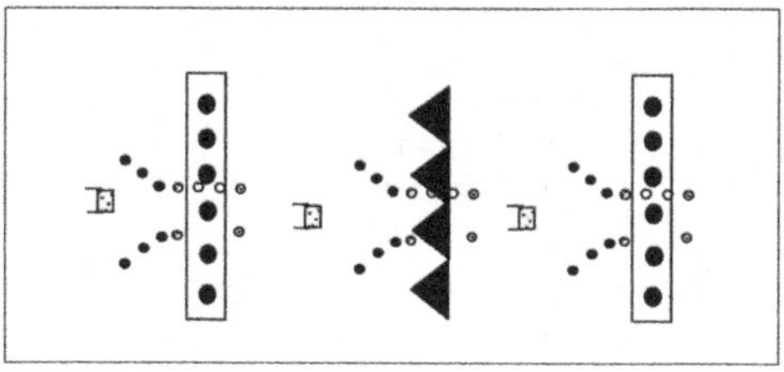

Figure 3-15. Marking a Complex Obstacle (Each Obstacle Independently Marked)

Figure 3-16. Marking a Complex Obstacle System

Chapter 4

Breaching Rehearsals

FM 101-5 gives a complete discussion of rehearsals, including types, techniques, and responsibilities. This chapter specifically discusses breaching rehearsals and considerations when conducting breaching rehearsals.

PLANNING

4-1. Rehearsals do not just happen; they must be well-planned. Successful rehearsals minimize the cost of time and maximize the understanding of the plan. Units may use a WO to establish responsibilities for preparing the rehearsal site. The WO can also identify battle drills that subordinate units should rehearse to prepare for the upcoming breaching operation. Units should include a complete rehearsal plan with an operation order (OPORD). The rehearsal plan may include a separate set of graphics if necessary.

TYPES

4-2. The five types of rehearsals are—

- Confirmation brief.
- Back brief.
- Combined-arms rehearsal.
- Support rehearsal.
- Battle drill or SOP rehearsal.

The combined-arms rehearsal and the battle drill or SOP rehearsal are most commonly used for breaching operations.

COMBINED-ARMS REHEARSAL

4-3. The maneuver unit HQ normally conducts the combined-arms rehearsal, and it is performed after subordinate units have issued their OPORD. The rehearsal ensures that subordinate plans are synchronized with those of other units and that the plans achieve the intent of the higher commander. A combined-arms rehearsal is particularly important when preparing for a complex breaching operation. It is very difficult to synchronize the actions of all the units involved in the operation.

4-4. Units should rehearse each phase of the operation, from the movement to and occupation of SBF positions through the passage of follow-on forces. The rehearsal must ensure that each unit knows how its tasks are integrated into the overall plan, in time and in space. For example, the support force must understand how long the breach force expects the reduction effort to take, so that the support force can ensure that its plan to suppress the enemy is adequate. At the same time, the breach force must understand its relative position to the support force when it is at the obstacle location. This ensures that the breach force, while moving to and reducing the obstacle, does not unnecessarily mask the fires of the support force.

4-5. At a minimum, the following should be addressed during a combined-arms rehearsal:

- The latest information concerning the enemy obstacle system.
- Actions that the higher HQ is taking to assist the breaching operation.
- The time available for emplacing smoke and activating CFZs (include location and duration).
- The criteria for lifting/shifting direct and indirect fires.
- The commitment criteria for the breach force.
- The method used to reduce the obstacle system.
- The marking to be used. One technique is to have a marked lane near the rehearsal site. All participants should view and drive/walk through the marked lane when arriving at or departing from the rehearsal site. The marking system should also be shown at support rehearsals conducted by combat service support (CSS) elements.
- The criteria and signals for the assault force to begin moving to created lanes.

BATTLE DRILL

4-6. A battle drill or an SOP rehearsal ensures that all participants understand techniques and procedures. These rehearsals are used by all echelons; but they are used most extensively by platoons, squads, and sections. For breaching operations, a battle drill or an SOP rehearsal can include lane-marking procedures or dismounted reduction techniques. Rehearsals should be conducted often to reinforce procedures that will be used during the breaching operation.

TECHNIQUES

4-7. Rehearsal techniques are limited only by the unit's resourcefulness. Generally, six techniques are used—

- Full dress.
- Reduced force.
- Terrain model.
- Sketch map.
- Terrain map.
- Radio.

Time and resource requirements for these techniques range from extensive to minimal. As the techniques are listed above, each successive one takes a decreasing amount of time and resources to prepare and conduct. Each rehearsal technique provides a different degree of understanding to participants. *Figure 4-1* shows the rehearsal techniques relative to time, resourcing, participation, and understanding.

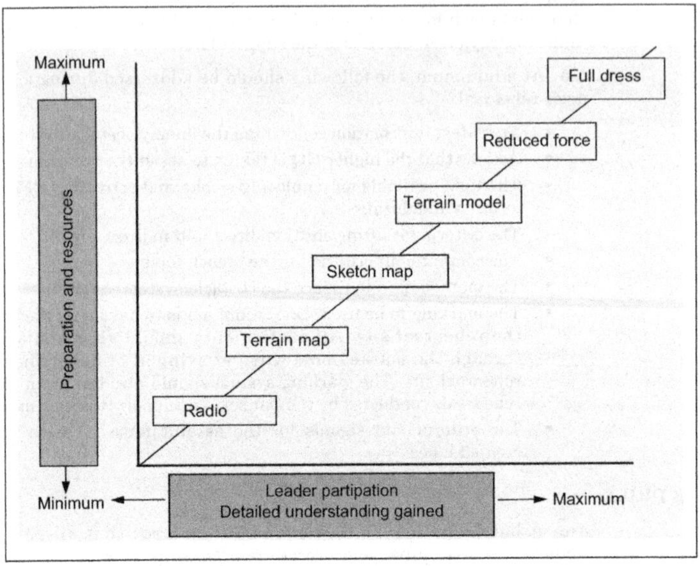

Figure 4-1. Rehearsal Techniques

4-8. When preparing for a breaching operation, use rehearsal techniques that provide the best understanding. Therefore, conduct a full-dress rehearsal if time and resources permit. Regardless of the rehearsal technique used, commanders must ensure that conditions (enemy situation, obstacle composition, visibility) are replicated in a realistic manner. For example, if a breach force is conducting a full-dress rehearsal, it should use the same obstacle system in the rehearsal as the one intelligence indicates that the enemy has emplaced. If intelligence is unavailable, the breach force should use the templated obstacle system. This requires the breach force to apply the same reduction techniques in the rehearsal that will be used during the operation.

4-9. Additionally, it is imperative that high standards be used during rehearsals. The breach force should rehearse proofing and marking obstacles. The assault force must rehearse all aspects of its assault, including the dismounted, ground fight if it is expected during the operation.

Appendix A

Metric Conversion Chart

This appendix complies with current Army directives which state that the metric system will be incorporated into all new publications. *Table A-1* is a conversion chart.

Table A-1. Metric Conversion Chart

US Units	Multiplied By	Metric Units
Feet	0.3048	Meters
Inches	2.54	Centimeters
Inches	0.0254	Meters
Inches	25.4001	Millimeters
Miles	1.6093	Kilometers
Pounds	453.59	Grams
Pounds	0.4536	Kilograms
Metric Units	**Multiplied By**	**US Units**
Centimeters	0.3937	Inches
Grams	0.03527	Ounces
Kilograms	2.2046	Pounds
Kilometers	0.62137	Miles
Meters	3.2808	Feet
Meters	39.37	Inches
Meters	1.0936	Yards
Millimeters	0.03937	Inches

Appendix B

Threat Mine Operations

This appendix is intended to complement the information presented in other manuals on threat obstacle tactics. Commanders should use this information to give added realism to unclassified staff and combined-arms team training. Obstacle employment norms can change with METT-TC factors for a given AO. Therefore, preoperational training on templating, intelligence, recon, and reduction procedures must be based on the best information available before deployment.

MINE OPERATIONS

B-1. Threat formations contain considerable organic minefield-emplacement capability. Threat rapid-mining capability presents a serious challenge to friendly maneuver.

B-2. Armored tracked mine layers known as *GMZs* (*Figure B-1*) dispense mines at a predetermined spacing of 5.5 meters. Three GMZs can lay a 1,200-meter, three-row minefield, containing 624 mines, in 26 minutes. (This does not include the 12- to 15-minute reload and travel times; travel and reload times increase during limited visibility.) Doctrinally, this minefield would be broken into several minefields, each 200 to 300 meters long. Mine-laying helicopters also support threat mine operations. The HIP and HIND-D helicopters carry two or three dispenser pods of AP or AT mines. Artillery-fired SCATMINEs can also support threat mine operations.

Figure B-1. GMZ Armored Tracked Mine Layer

B-3. Threat armies use obstacles extensively throughout the depth of their defense, and their tactics are chosen well. A threat rapidly emplaced minefield consists of three or four 200- to 300-meter rows, spaced 20 to 40 meters apart, with mines spaced 4 to 6 meters apart.

B-4. *Table B-1* provides detailed information on standard threat AT and AP minefields. Terrain and tactical situations dictate the actual dimensions and distances of minefields.

Table B-1. Normal Parameters for Threat-Style Minefields

AT Minefields	
Front (situation-dependent)	200 to 300 meters
Depth	40 to 120 meters
Number of rows	3 or 4
Distance between rows	20 to 40 meters
Distance between mines	4 to 6 meters for antitrack mines; 9 to 12 meters for antihull mines
Outlay, normal	550 to 750 antitrack mines per kilometer; 300 to 400 antihull mines per kilometer
Outlay, increased effect	1,000+ antitrack mines per kilometer; 500+ antihull mines per kilometer
Probability of destruction	57 percent for antitrack mines (750 per kilometer); 85 percent for antihull mines (400 per kilometer)
AP Minefields	
Front (situation-dependent)	30 to 300 meters
Depth	10 to 150 meters
Distance between rows	5+ meters for blast mines; 25 to 50 meters for fragmentation mines
Distance between mines	1 meter for blast mines; 50 meters (or twice the lethal radius of fragmentation) for fragmentation mines
Outlay, normal	2,000 to 3,000 HE/blast mines per kilometer; 100 to 300 fragmentation mines per kilometer
Outlay, increased effect	2 to 3 times the normal outlay
Probability of destruction	15 to 20 percent for HE/blast mines (2,000 per kilometer); 10 to 15 percent for fragmentation mines (100 per kilometer)

B-5. *Figure B-2* shows a standard rapidly emplaced minefield. The threat army typically uses such a minefield when they are in a hasty defense (offense is temporarily stalled).

B-6. *Figure B-3* shows a standard antitrack minefield.

B-7. *Figure B-4*, page B-4, shows a standard antihull minefield.

B-8. *Figure B-5*, page B-4, shows a standard AP minefield.

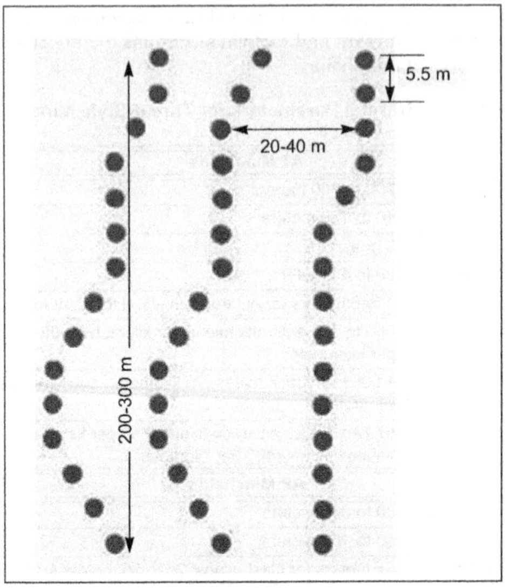

Figure B-2. Threat-Style Rapidly Emplaced Minefield

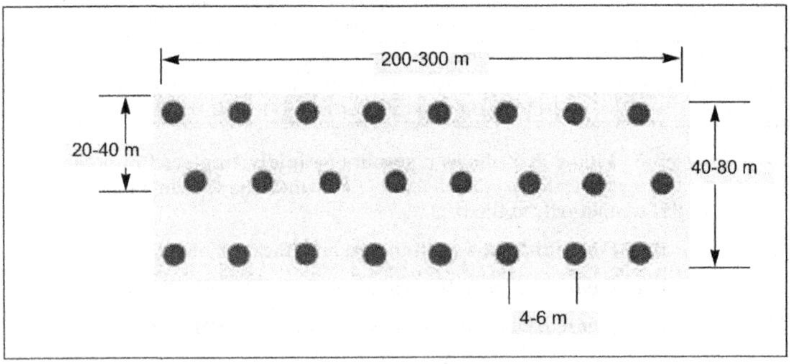

Figure B-3. Threat-Style Antitrack Minefield

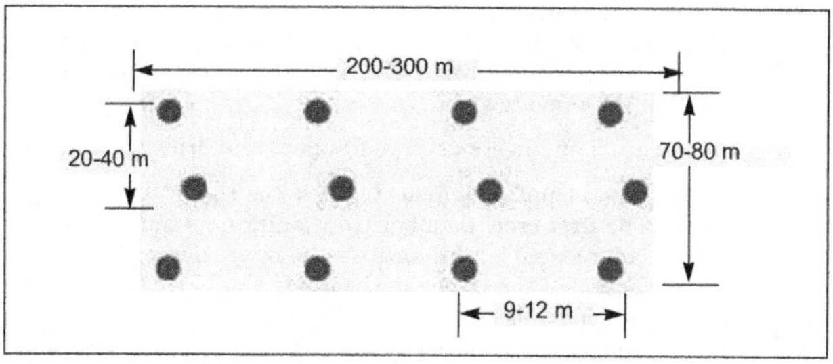

Figure B-4. Threat-Style Antihull Minefield

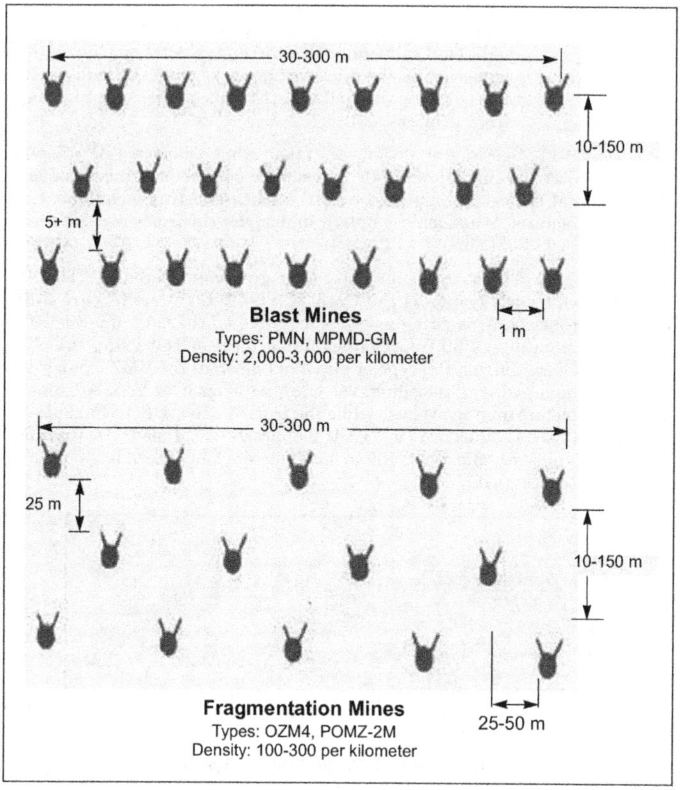

Figure B-5. Threat-Style AP Minefield

B-9. Threat armies also emplace mixed minefields. They are not the same as US mixed minefields. Threat armies normally emplace three rows of AT mines, then several rows of AP mines. AT and AP mines are not normally mixed in the same row.

B-10. Threat engineers use two fundamental drills to emplace mines:

- When emplacing armed mines, the threat uses a crew of five sappers. The first crew member (the senior man and operator) is in the mine layer's seat and monitors the operation of the mine layer and the motion of the mines in the guide chute. He also sets the mine spacing and controls the actions of the GMZ. The second and third members take the mines out of the containers and place them in the intake chute at intervals between the guide tray's drive chain. The GMZ driver steers the vehicle along the indicated route at the established speed.
- When emplacing unarmed mines, the threat uses two or three additional sappers to arm the mines. After emplacing the mines, one sapper trails the mine layer, marks emplaced mines with pennants, and partially camouflages the mines. The remaining sapper(s) then arm the mines.

B-11. Special precautions are taken when emplacing AP minefields. Threat doctrine only allows PMN mines to be surface-laid from mine layers. POMZ-2M mines are emplaced with the truck-and-tray technique. Extra effort is required to assemble, emplace, and deploy the trip wire and to camouflage the POMZ-2M mine.

B-12. Threat forces may also have ground-emplaced SCATMINE capability. One such system is the UMZ SCATMINE system (*Figure B-6*). The UMZ consists of six firing modules mounted on the back of a Zil-131 truck. Each module has 30 firing tubes, for a total of 180 firing tubes per system. Depending on the type of minefield desired, the UMZ can lay 180 to 11,520 mines without reloading. The UMZ can launch an AT or AP minefield 30 to 60 meters from the vehicle while the truck is driving 10 to 40 kilometers per hour (kph). It takes two men 1½ to 2 hours to reload the UMZ. One UMZ can lay a three-row minefield, 150 to 1,500 meters long, depending on the type of mine that is used.

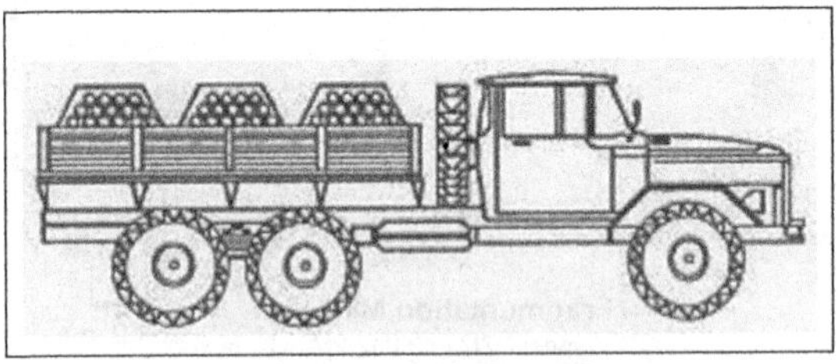

Figure B-6. UMZ SCATMINE System

B-13. For hand-emplaced SCATMINEs, there is a man-portable SCATMINE dispenser. The PKM weighs 2.63 kilograms (without the mine canister) and consists of a single launch tube with a base mount, a blasting machine, and a reel of electric ignition wire. The operator loads a propelling charge and a mine canister into the launch tube and mounts the tube on the edge of a trench or firing parapet. He then aims the tube, connects the ignition wire to the tube, and moves to a safe distance. At an initiating point, the operator connects the ignition wire to the blasting machine and initiates the system. The PKM propels the canister 30 to 100 meters, depending on the type of mine. It lays an AP minefield that is 10 by 20 meters (POM-1S or PFM-1S mine canister) or 10 by 40 meters (POM-2S mine canister). It takes a trained operator 5 minutes to set up the PKM and create a minefield. The PKM can also be used to launch the PTM-1S and PTM-3 AT mine canisters.

B-14. Threat forces use the PKM to lay minefields that protect subunit positions and flanks and the boundaries between subunits. PKM-laid minefields also cover firing lines and gaps in combat formations. The PKM can quickly close breaches in existing minefields and increase the density of mines on armor AAs.

B-15. The type and complexity of an obstacle depend on the installing unit. Maneuver and artillery soldiers usually install simple single-system minefields that are protective in nature. Engineer soldiers install complex obstacles that can include AHDs. Engineer obstacle placement is usually equipment-intensive. Threat engineer effort generally concentrates on tactical obstacles unless maneuver soldiers are unable to employ the necessary protective obstacles. Threat units continue to improve obstacles, supporting their positions by marking the friendly side of the obstacles, burying mines, and adding AHDs.

FOREIGN-MINE DATA

B-16. *Tables B-2 through B-7*, pages B-7 through B-11, provide details on foreign AT and AP mines and mine delivery systems.

Table B-2. Foreign Track-Width AT Mines

Origin	Mechanically Planted	Scatterable	Metallic	Nonmetallic	Simple Pressure	Double Impulse	Blast Hardened	AHD
Russia	TM-46 TM-57 TM-62	PTM-1	TM-46 TM-57 TM-62	TM-62B TM-62D TM-62P	TM-46 TM-57 TM-62		TM-46 TM-57	TMN-46
China	Type 72	Type 84B	Type 72	Type 72	Type 84B	Type 72	Type 72	
Yug	TMM-1		TMM-1	TMA-1 TMA-2 TMA-3 TMA-4 TMA-5	TMA-1 TMA-2 TMA-5		TMA-3 TMA-4	TMM-1 TMA-1 TMA-2 TMA-3 TMA-4 TMA-5
Czech	PT-MI-K PT-MI-BA III		PT-MI-K PT-MI-D	PT-MI-BA II PT-MI-BA III	PT-MI-BA III PT-MI-D		PT-MI-K PT-MI-BA II	PT-MI-K PT-MI-D
Italy	VS-1.6 VS-2.2 TC-.6/6 SB-81	VS-1.6 SB-81		VS-1.6 VS-2.2 TC-.6/6 SB-81			VS-1.6 VS-2.2 TC-3.6 TC-6 SB-81	VS-1.6 VS-2.2
France	M-51 M-52			M-51 M-52	M-51 M-52			M-51
UK	L9 Barmine MK-7		MK-7	L9 Barmine	L9 Barmine MK-7	L9 Barmine MK-7		L9 Barmine
Germany	PM-60 DM-11		DM-21	PM-60 DM-11	PM-60 DM-11 DM-21			PM-60 DM-11 DM-21

Table B-3. Foreign Full-Width AT Mines

Origin	Mechanically Planted	Scatterable	Tilt Rod	Magnetic Influence	Seismic Influence	Coupled	AHD
Russia	TM-62 TM-72 TM-89	PTM-3	TMK-2	TM-62 TM-72 TM-89	TM-62 TM-72 TM-89		TM-62 TM-72 TM-89
China			Type 84C	Type 84A			Type 84A
Yug	TMRP-6	KB-2	TMRP-6	TMRP-6 KB-2			TMRP-6 KB-2
Czech			PT-MI-P				
France	HPD series	Lance	M-51 M-52 Model 1954	HPD series		Model 1953	HPD series M-51 Lance
Hungary	UKB-63		UKB-63				UKB-63
Italy	VS-HCT VS-HCT-2 SB-MV	VS-SATM1 VS-HCT		VS-HCT VS-HCT-2 SB-MV VS-SATM1		Fuse option	VS-HCT VS-HCT-2 SB-MV VS-SATM1
Sweden	FFV-028			FFV-028		Fuse 15	FFV-028
UK	L9 Barmine	HB-876	L9 Barmine MK-7	L9 Barmine HB-876			L9 Barmine MK-7 HB-876
Germany	DM-31	AT-2 MIFF	AT-2	DM-31 MIFF			DM-31 AT-2 MIFF
Romania						MC-71	

Table B-4. Foreign Pressure-Fused AP Mines

Origin	Scatterable	Low Metal	Simple Pressure	Small Diameter	Low-Operating Threshold	Pronged Fuse	Blast Hardened	AHD
Russia	PFM-1		PMN PMD-6		PMN PMD-6 PFM-1	OZM series	PMN-2	
China	GLD-111	Type 72A GLD-111	Type 72A Type 58 GLD-111	GLD-111	Type 72 Type 58 GLD-111			Type 72A Type 72B
Yug		PMB-1 PMB-2 PMB-3	PMA-1 PMD-1		PMA-1 PMA-3 PMD-1	PROM-1 PROM-2 PMR-3	PMA-2 PMA-3	
Czech			PP-MI-D PP-MI-BA		PP-MI-D	PP-MI-SR PP-MI-SR II		
Italy	TS-50 VS-50 SB-33 VS-MK2	TS-50 SB-33 VS-MK2				Valmara 59 Valmara 69	TS-50 VS-50 SB-33 VS-MK2	VS-50 SB-33 VS-MK2
France		M-59		M-59	M-59		M-59	
UK	Ranger		Ranger	Ranger				
Germany		DM-11	PPM-2		DM-11		DM-11	

Table B-5. Foreign Side-Attack AT Mines

Origin	Command	Break Wire	Influence	AHD
Russia	MON-100/200/500, TM-83	LMG	TM-73/83	
Yug		TMRP-6, COBRA		
France	MIACAH F1	MIACAH F1	MIACAH F1	
UK	L14	L14	AJAX	AJAX
Germany			DM-12	
Sweden		FFV-016, AT-4		

Table B-6. Foreign Trip-Wire/Break-Wire-Fused AP Mines

Origin	Scatterable	Trip Wire	Break Wire
Russia	POM-2	POMZ-2, OZM series, MON series	OZM series, MON series
China		Type 58, Type 59, Type 66, Type 69	
Yug		PROM-1/2/KD, PMR-2/3/4, MRUD	PROM-1/2/KD, PMR-2/3/4, MRUD
Czech		PP-MI-SR, PP-MI-SR II, PP-MI-SK	
Italy	VS-SAPFM3	Valmara 59/69, BM-85, P-25, P-40, VS-DAFM1	
France		M1951/1955, MAPED F1	M1951/1955
UK		PAD	
Germany		DM-31, SM-70	

Table B-7. Foreign Mine Delivery Systems

Origin	System	Type[1]	Range	Mine Payload	Density[2]	Remarks
Russia	BM-21	R	20 km	PTM 1/3, POM-2	1.5	1,000 x 500 m per battery
	BM-22	R	35 km	PTM 1/3, POM-2	1.5	
	BKPI	F, H	NA	PTM 1/3, PFM-1, POM-2	1.5	Lays relatively narrow strips
	KSMU	F	NA	PTM 1/3, PFM-1, POM-2	1.5	
	UMZ	V	NA	PTM 1/3, POM-2	1.5	
	GMZ/PMZ-4	MP	NA	Varies	0.4 to 1.0	3 systems can emplace 1,200 m with 624 mines in 26 minutes
	VMR-1/2	H	NA	Varies	0.4 to 1.0	3 Mi-8s can lay 1,200 to 3,000 m with 1,200 mines
China	Type 83	R	15 km	Type 84	0.5	900 x 800 m per battery
Yug	M87	R	50 km	KB-2	0.5	900 x 500 m per battery

Table B-7. Foreign Mine Delivery Systems (continued)

Origin	System	Type[1]	Range	Mine Payload	Density[2]	Remarks
Italy	FIROS 25	R	22 km	VS-HCL, VS-Mk, VS-SAPFM 3	0.5	
	FIROS 30	R	35 km	VS-HCL, VS-Mk, VS-SAPFM 3	0.5	
	DAT	H	NA	Varies	1.2	Lays relatively narrow strips of up to 320 AT mines
	Istrice	V	NA	SATM1, VS-Mk 2 EL, VS-SAPFM 3	Max 1.6	Typically 360 x 140 m
	ST-AT/V	MP	NA	Varies	0.3 to 1.5	240 mines per hour
	Grillo 90	MD	NA	SATM1, VS-Mk 2 EL	0.5	5 AT or 15 AP mines per dispenser
Sweden	FFV-5821	MP	NA	FFV-028	0.4 to 1.4	300 buried mines per hour or 500 surface-laid mines per hour
France	Minotaur	V	NA	Lance	0.7	1,200 x 600 m with 800 mines
	EBG	V	NA	Lance	1.5	60 x 600 m with 40 mines
	155-mm How	A	18 km	Lance		8 mines per round
	Matenin	MP	NA	HPD series	0.4 to 1.4	500 mines per hour
UK	JP-233	F	NA	HB 876		430 mines per Tornado aircraft
	Barmine layer	MP	NA	L9 Barmine	0.4 to 0.9	Lays up to 700 mines per hour
	Ranger	V	NA	L10		1,296 mines per dispenser
Germany	Skorpion	V, H	NA	AT-2	0.4	1,500 x 50 m with 600 mines
	MW-1	F	NA	MIFF, MUSPA	0.5	55 to 500 m wide and 200 to 2,500 m
	LARS	R	14 km	AT-2	0.5	1,836 AT mines per launcher, 1,000 x 500 m per battery
	MARS	R	30 km	AT-2	0.3	1,000 x 500 m with 336 AT mines per launcher
	FFV-5821	MP	NA	DM-31	0.4 to 1.4	300 buried mines per hour or 500 surface-laid mines per hour

[1] A = artillery, H = helicopter, F = fixed-wing aircraft, R = rocket, V= vehicle dispensed, MD = man-portable dispenser, MP = mechanical planter

[2] AT mines per meter of front

Appendix C

Obstacle-Reduction Techniques

Obstacle reduction is the physical creation of a lane through an obstacle. It is a fundamental of breaching operations as discussed in earlier chapters. A number of tasks (detecting, reporting, reducing, proofing, and marking) directly support or are included in obstacle reduction.

DETECTING

C-1. Detection is the actual confirmation of the location of obstacles. It may be accomplished through recon, or it may be unintentional (such as a vehicle running into a mine). Detection is used in conjunction with intelligence-gathering operations, bypass recons, and breaching and clearing operations. There are four types of detection methods—visual, physical (probing), electronic, and mechanical.

VISUAL

C-2. Visual detection is part of all combat operations. Personnel visually inspect the terrain for the following obstacle indicators:

- Trip wires.
- Signs of road repair (such as new fill or paving, road patches, ditching, and culvert work).
- Signs placed on trees, posts, or stakes. Threat forces mark their minefields to protect their own forces.
- Dead animals.
- Damaged vehicles.
- Disturbances in previous tire tracks or tracks that stop unexplainably.
- Wires leading away from the side of the road. They may be firing wires that are partially buried.
- Odd features in the ground or patterns that are not present in nature. Plant growth may wilt or change color, rain may wash away some of the cover, the cover may sink or crack around the edges, or the material covering the mines may look like mounds of dirt.
- Civilians who may know where mines or booby traps are located in the residential area. Civilians staying away from certain places or out of certain buildings are good indications of the presence of mines or booby traps. Question civilians to determine the exact locations.
- Pieces of wood or other debris on a road. They may be indicative of pressure or pressure-release firing devices (FDs). These devices may be on the surface or partially buried.

- Patterns of objects that could be used as a sighting line. An enemy can use mines that are fired by command, so road shoulders and areas close to the objects should be searched.
- Berms. They may indicate the presence of an AT ditch.

PHYSICAL

C-3. Physical detection (probing) is very time-consuming and is used primarily for mine-clearing operations and self-extraction. Detection of mines by visual or electronic methods should be confirmed by probing. Use the following procedures and techniques when probing for mines:

- Roll up your sleeves and remove your jewelry to increase sensitivity. Wear a Kevlar™ helmet, with the chin strap buckled, and a protective fragmentation vest.
- Stay close to the ground and move in a prone position to reduce the effects of an accidental blast. When moving into a prone position—
  - Squat down without touching your knees to the ground.
  - Scan forward up to 2 meters and to the sides up to 3 meters for mine indicators.
  - Probe the area around your feet and as far forward as possible.
  - Kneel on the ground after the area is found to be clear, and continue probing forward until you are in a prone position.
- Use sight and touch to detect trip wires, fuses, and pressure prongs.
- Use a slender, nonmetallic object as a probe.
- Probe every 5 centimeters across a 1-meter front.
- Push the probe into the ground gently at an angle that is less than 45 degrees.

DANGER
Use extreme caution when probing. If the probe is pushed straight down, its tip may detonate a pressure fuse.

- Apply just enough pressure on the probe to sink it slowly into the ground.
- Pick the soil away carefully with the tip of the probe. Remove the loose dirt by hand if the probe encounters resistance and does not go into the ground freely. Take care to prevent detonating the mine.
- Stop probing and use two fingers from each hand to carefully remove the surrounding soil when a solid object is touched and identify the object.
- Remove enough soil to show the mine type (if the object is a mine) and mark its location. Do not attempt to remove or disarm the mine. Use explosives to destroy detected mines in place or use a grappling hook and rope to cause mines to self-detonate. Do not use metal grappling hooks on magnetic-fused mines.

C-4. Probing is extremely stressful and tedious. The senior leader must set a limit to the time a prober will actually probe. To determine a reasonable time, the leader must consider METT-TC factors, weather conditions, the threat level, the unit's stress level, and the prober's fatigue level and state of mind. As a rule, 20 to 30 minutes is the maximum amount of time that an individual can probe effectively.

ELECTRONIC

C-5. Electronic detection is effective for locating mines, but this method is time-consuming and exposes personnel to enemy fire. In addition, suspected mines must be confirmed by probing.

C-6. The AN/PSS-12 mine detector (*Figure C-1*) can only detect metal, but most mines have metal components in their design. The detector can locate and identify plastic or wooden mines by a slight metallic signature. Employment and operation procedures for the AN/PSS-12 are discussed in *FM 20-32*, and technical data is available in *Technical Manual (TM) 5-6665-298-10*. The detector is handheld and identifies suspected mines by an audio signal in the headphones.

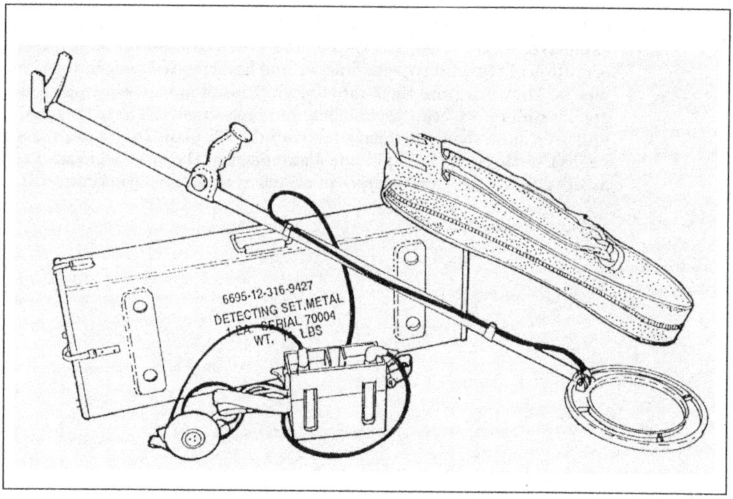

Figure C-1. AN/PSS-12 Mine Detector

C-7. As in probing, consideration must be taken for the maximum amount of time an individual can operate the detector. As a rule, 20 to 30 minutes is the maximum amount of time an individual can use the detector effectively.

MECHANICAL

C-8. The track-width mine roller is a mechanical minefield-detection system. It is most effectively deployed to lead columns on route movement, but it can be used to precede tactical formations. In column movement, unit vehicles travel a narrow path and one or two mine rollers can effectively detect mines in the path. Mine rollers can also be used to detect minefields in front of deployed tactical formations, although more than one roller is required for a good probability of detection.

REPORTING

C-9. Intelligence concerning enemy minefields is reported by the fastest means available. Spot reports (SPOTREPs) are the tactical commander's most common source of minefield intelligence. They originate from patrols that have been sent on specific minefield recon missions or from units that have discovered mine information in the course of their normal operations. The information is transmitted to higher HQ.

REDUCING

C-10. Obstacle-reduction and clearing equipment is broken down into explosive, mechanical, electronic, and manual. Combat engineers and the operators of reduction assets practice and become proficient in these reduction means. They integrate them into breach drills of units they support. The team applies different tactics, techniques, and procedures (TTP) to breach drills and prepares and rehearses them as part of the TF plan. *Tables C-1 through C-4*, pages C-5 through C-11, provide a summary of the effectiveness of reduction assets against the various types of obstacles that may be encountered.

Table C-1. Mechanical Reduction Assets Versus Threat Obstacles

Asset	Track-Width AT Mine	Full-Width AT Mine	Side-Atk AT Mine	Cmd-Det Blast Mine	AT Ditch and Other Gaps	Wire Obstacles	Remarks
IDA	Is more than 90 percent effective to 30.4 cm deep at speeds less than 10 kph (soil dependent)	Is effective against magnetic-influence mines; is more than 99 percent effective against tilt-rod mines	Do not use	Do not use	May be effective at filling in some shallow gaps or trenches	Do not use	The MCB provides almost 100 percent clearance of simple pressure-fused AT mines when used with the MICLIC. The MCB has minimal impact on the M1's maneuverability and speed when not plowing. The main gun should be traversed to the side to prevent damage in case the mine detonates when plowing. The point of breach should be relatively flat and free of large rocks and other obstructions. Plowing operations should begin about 100 m short of the suspected minefield and continue 100 m past its suspected end (typically 2 to 2½ minutes to cover 300 to 400 m overall). Plowing operations should be performed in a straight line to keep the vehicle out of uncleared ground. Fifteen percent of pressure-fused AT-mine encounters across the MCB's width can be expected to detonate under the skid shoe, probably disabling the blade. The MCB is vulnerable to coupled mines and may be stopped by log-hurdle obstacles laid within a minefield as part of a complex obstacle. A second plow should not be used to clear the center of the lane because this will simply push mines into the area cleared by the first plow.

Table C-1. Mechanical Reduction Assets Versus Threat Obstacles (continued)

Asset	Track-Width AT Mine	Full-Width AT Mine	Side-Atk AT Mine	Cmd-Det Blast Mine	AT Ditch and Other Gaps	Wire Obstacles	Remarks
MCR with IDA	Is more than 90 percent effective to 30.4 cm deep at speeds less than 10 kph (soil dependent)	Is effective against magnetic-influence mines; is more than 99 percent effective against tilt-rod mines	Do not use	Do not use	Do not use	Is effective against barbed wire	Multiple rollers are required to provide the overlap necessary for proofing the route's entire width when the MCR is used for proofing during route-clearance operations. Vehicle speed is limited to 5 to 15 kph. The main gun should be traversed to the side to prevent damage in case the mine detonates when rolling a suspected mined area. The IDA may be damaged by detonated mines.
AVLB	NA	NA	NA	NA	Can cross up to MLC 60 (70 if improved) traffic over gaps up to 17.4 m (unprepared abutments) and 18.3 m (prepared abutments)	NA	Is carried on a launcher (modified M48A5 or M60A1 chassis) Weighs 15 tons (bridge only) Can be launched in 2 to 5 minutes
Wolverine	NA	NA	NA	NA	Can cross MLC 70 traffic over gaps up to 24 m	NA	Is carried on a launcher (modified M1-series Abrams tank chassis) Weighs 12.5 tons (bridge only) Can be launched in less than 5 minutes
ACE	NA	NA	NA	NA	Is very effective	NA	

Table C-1. Mechanical Reduction Assets Versus Threat Obstacles (continued)

Asset	Track-Width AT Mine	Full-Width AT Mine	Side-Atk AT Mine	Cmd-Det Blast Mine	AT Ditch and Other Gaps	Wire Obstacles	Remarks
MiniFlail	Is effective against simple pressure-fused AP mines Is less effective against blast-hardened AP mines	Is effective	Is ineffective	Is armored against frag-menting threats Is vulner-able to command-detonated blast mines	NA	Do not use	AT mines can destroy the MiniFlail. Commanders must weigh the use of the MiniFlail against other reduction assets when conducting minefield-clearance operations. Can clear at the rate of about 1,500 square meters per hour

Table C-2. Manual Reduction Assets Versus Threat Obstacles

Asset	Pressure-Fused AP Mine	Trip-Wire/ Break-Wire-Fused AP Mine	Influence-Fused AP Mine	Cmd-Det Blast Mine	AT Ditch and Other Gaps	Wire Obstacles	Remarks
APOBS	Effectively clears 95 percent, including blast-resistant types	Effectively cuts 99 percent of trip wires in its path	Is ineffective	Is ineffective	NA	Clears a footpath	Provides a 0.6- x 45-m footpath from a 25-m standoff Can be deployed by two men in 2 minutes Consists of two manpack units weighing about 55 lbs each
Bangalore torpedo	Effectively clears 85 percent of simple pressure-fused mines; is less effective against blast-hardened mines	Effectively cuts 99 percent of trip wires in its path	Is ineffective	Is ineffective	NA	Clears a footpath	Provides a 0.6- x 15-m footpath Can be deployed by a squad in 5 minutes Consists of ten 1.5-m-long sections weighing 13 lbs each Is ineffective against AT mines
WLGH	Is ineffective	Effectively clears 99 percent of trip wires/break wires in a single pass	Is ineffective; however, it may snag the wires between the control unit and its mines	Is ineffective; however, it may snag the wires between the observer and his mines	NA	Is somewhat effective at clearing trip-wire-fused mines integrated with wire obstacles	Has a range of 80 to 100 m when launched Can be thrown 15 to 20 m when hand thrown Must consider the lethal range/radius (up to 90 m in some cases) of fragmenting AP mines that may be encountered. Select a covered position or move back an adequate distance before pulling the grapnel. Helps increase confidence by grappling an area several times Is hampered by thick vegetation/overhead obstructions that may reduce the WLGH's range/effectiveness

Table C-2. Manual Reduction Assets Versus Threat Obstacles (continued)

Asset	Pressure-Fused AP Mine	Trip-Wire/ Break-Wire-Fused AP Mine	Influence-Fused AP Mine	Cmd-Det Blast Mine	AT Ditch and Other Gaps	Wire Obstacles	Remarks
Probe	Is effective at locating all buried mines; however, extreme caution should be used when probing for small mines (less than 7.6 cm in dia) and sensitive mines (more than 10 lb of actuating force)	Is ineffective; the soldier must rely on clearance by other means and visual detection	Do not use	Ensure that observer is neutralized before approaching	NA	Is ineffective	
Wire cutter	Is ineffective	Do not cut a slack wire until both ends have been checked if break-wire-fused mines may be present. Do not cut taut trip wires until both ends have been checked for tension-release FDs	Do not use	Ensure that observer is neutralized before approaching	NA	Is effective; sound can be minimized by wrapping the wire with a cloth before cutting it	
Gunfire	Is effective against surface-laid munitions	Is effective against surface-laid munitions	Do not use	Is effective against surface-laid munitions	NA	Is ineffective	Is used as a last resort Is a relatively slow method for clearing visually detected ordnance Ensures protection of weapons operator Works better with larger caliber weapons (up to 25 mm); incendiary ammo works best May be the only option open to artillery, MP, and transportation units for dealing with SCATMINEs in the absence of engineers and reduction assets

Table C-3. Dismounted Breaching Assets Versus Threat Obstacles

Asset	Pressure-Fused AP Mine	Trip-Wire/ Break-Wire-Fused AP Mine	Influence-Fused AP Mine	Cmd-Det Blast Mine	AT Ditch and Other Gaps	Wire Obstacles	Remarks
Demolitions	Place at least a 1-lb charge within 15.2 cm of simple pressure-fused mines Ensure that the charge is placed within 2.54 cm of blast-hardened mines	Place at least a 1-lb charge within 15.2 cm of the mine after the mine at the end of a trip wire has been located Can use elevated charges if necessary against the claymore and stake-type mines	Do not use	Ensure that observer is neutralized before approaching Can use elevated charges if necessary against claymore mines	NA	Is effective when used to make an improvised bangalore torpedo	
Blast overshoes	May be effective against very small (up to 1 oz of explosive) mines Is ineffective against mines with more than 1-oz explosive weight	Can reduce casualty severity against fragmenting mines	Can reduce casualty severity against fragmenting mines	Ensure that observer is neutralized before approaching Can reduce casualty severity	NA	Is ineffective	
IBASIC	Will protect personnel near the individual who detonates a pressure-fused AP mine	Can reduce casualty severity against fragmenting mines	Can reduce casualty severity against fragmenting mines	Ensure that observer is neutralized before approaching Can reduce casualty severity	NA	Is ineffective	

Table C-4. Electronic Detection Assets Versus Threat Obstacles

Asset	Pressure-Fused AP Mine	Trip-Wire/ Break-Wire-Fused AP Mine	Influence-Fused AP Mine	Cmd-Det Blast Mine	AT Ditch and Other Gaps	Wire Obstacles	Remarks
AN/PSS-12	Is very effective at finding metallic mines Is less effective against low-metal mines	Is effective at finding metallic trip wires and the fragmenting mines to which they are normally attached Cannot detect monofilament fishing line used for trip wires; it is undetectable by the AN/PSS-12. Visual detection and other neutralization means should be used first	Do not use	Ensure that observer is neutralized before approaching	NA	Is ineffective	Is less effective against low-metal mines and in areas contaminated with metal fragments (artillery, mortar, small arms) or areas that have heavily mineralized soils

EXPLOSIVE

C-11. The different types of explosive minefield-reduction and clearing equipment are discussed below.

M58A4 Mine-Clearing Line Charge

C-12. The MICLIC (*Figure C-2*) is a rocket-propelled, explosive line charge. It is used to reduce minefields that contain single-impulse, pressure-activated AT mines and mechanically activated AP mines. It clears a 14- by 100-meter path. The MICLIC has a 62-meter standoff distance from the launcher to the detonation point. The MICLIC's effectiveness is limited against prong AP mines, magnetically activated mines (including some SCATMINEs), top-attack mines, side-attack mines, and mines containing multiple-impulse or delay-time fuses. It also has little effect on other obstacles, such as log and concrete barriers, antivehicular ditches, and walls. The shock effect and the psychological impact of the detonation make the MICLIC a useful weapon in a close fight or in urban areas.

Figure C-2. MICLIC

NOTE: See *Figures C-3* and *C-4* for information on the MICLIC's surface danger zone.

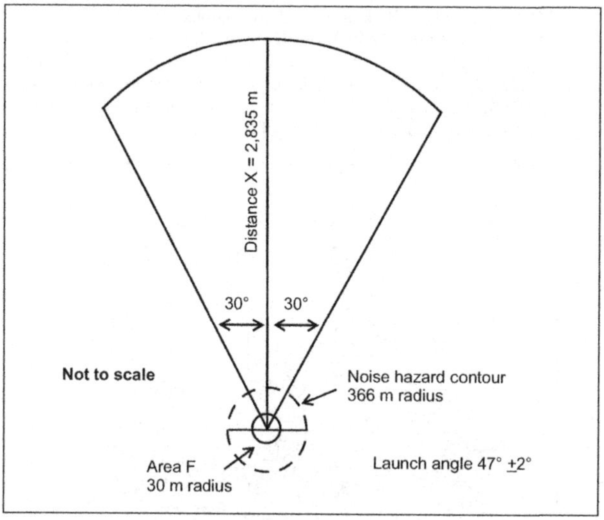

Figure C-3. Surface Danger Zone for Firing the MICLIC

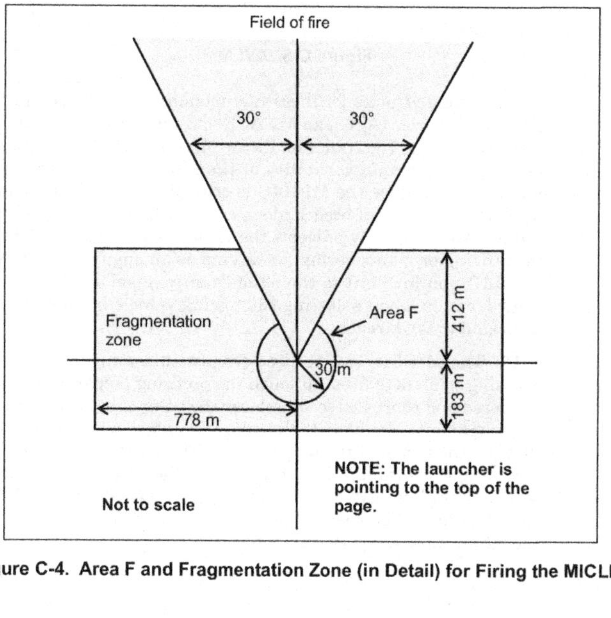

Figure C-4. Area F and Fragmentation Zone (in Detail) for Firing the MICLIC

C-13. The MICLIC is mounted on a rubber-tired trailer, or two MICLICs can be mounted on an AVLB, with the bridge downloaded using a fabricated I-beam frame. Procedures for mounting the MICLIC on the AVLB are outlined in *TM 9-1375-215-14&P*. This is called an armored vehicle-launched MICLIC (AVLM) (*Figure C-5*) and may be the preferred system because no trailer is involved to hinder the mobility of the vehicle.

Figure C-5. AVLM

C-14. Towing vehicles for the trailer-mounted MICLIC are an M113 armored personnel carrier (APC), an M9 ACE, a 5-ton wheeled vehicle, and a 2½-ton wheeled vehicle. The trailer limits the MICLIC's mobility in rough terrain and degrades the maneuverability of the towing vehicle, thereby increasing vulnerability. Since the MICLIC is critical to the breaching operation, it is moved to the point of breach along easily trafficable, covered, and concealed routes. This effectively prevents the towing vehicle from performing any other task (firing or maneuvering) or serving as an engineer squad vehicle unless MICLIC employment is the squad's only mission. This is an important consideration when selecting the towing vehicle because it must be solely dedicated to the mission.

C-15. The MICLIC can be fired from within an armored vehicle without exposing soldiers to fires, although the prefiring preparations must be done at a covered and concealed location near the point of breach. The initiating cable is brought into the vehicle through the hatch, which must be left ajar, or through the portal of the periscope, which has been removed. Therefore, the crew is not afforded NBC protection. When the MICLIC is fired from a wheeled vehicle, the crew must move to a covered position outside the backblast area. The special-purpose cable on the firing control switch is long enough to allow adequate standoff.

C-16. The vehicle operator must be proficient in all aspects of preparing and deploying the MICLIC, including the critical aspect of selecting the optimum point of breach. Although the engineer platoon leader or the breach force commander directs the operator to the point of breach, ensuring that the operator can independently accomplish the task will simplify the operation and greatly enhance its likelihood of success. The towing vehicle and the operator must be selected well in advance and be dedicated solely to the task. The operator must be included in all rehearsals and planning sessions and, if possible, during leader recons. The MICLIC crew and the breaching force must ensure that misfire/malfunction drills are addressed during the unit rehearsal.

C-17. Each MICLIC trailer transports and fires one charge, and then it must be reloaded. The AVLM can fire both MICLICs before reloading. The loaded-charge container weighs 1,283 kilograms, so a lifting device such as a 5-ton wrecker or a heavy expanded mobility tactical truck (HEMTT) is needed. Reloading, which can be done by an experienced crew in about 20 minutes, entails loading a rocket on the rail and lifting a new charge container onto the launcher. The reloading operation must be done in a covered and concealed location.

C-18. The exact limits and depth of an enemy minefield are seldom known before a breach. This is particularly true when the situation is unclear and the minefield is encountered simultaneously with enemy contact. The first and only indication that a unit is in a minefield may be when a vehicle encounters a mine. The minefield's leading edge may still be an uncertainty because the vehicle could have hit a mine in an interior row. The number of MICLICs needed to clear a single lane through a minefield—

- Less than 100 meters deep is one (see *Figure C-6*, page C-16). The minefield's leading edge is identified and, if possible, confirmed by recon. The MICLIC is deployed from a minimum standoff distance of 62 meters from the minefield's leading edge.
- More than 100 meters deep or of uncertain depth is two or more (see *Figure C-7*, page C-16). If the leading edge cannot be identified, the MICLIC is deployed 100 meters from the possible edge or stricken vehicle. When the first MICLIC is detonated, a second MICLIC moves 25 meters into the first MICLIC's path and fires its charge. This extends the lane an additional 87 meters. Additional MICLICs are used for minefields of extreme depth, and each one moves down the lane 25 meters into the path created by the previous charge.

C-19. The neutralization of mines by blast depends on the peak pressure and the impulse. For the MICLIC, the impulse is at a maximum of 3 meters from the line charge (on both sides) and decreases the closer it gets toward the line charge to a minimum of 1 meter from the line charge. This decrease on impulse causes a *skip zone* (see *Figure C-8*, page C-17). This does not mean that neutralization is equal to zero percent; it means that it is not equal to 100 percent. Mines that are buried deeper than 10 centimeters and located 1 to 2 meters from the line charge have a high probability of not being neutralized.

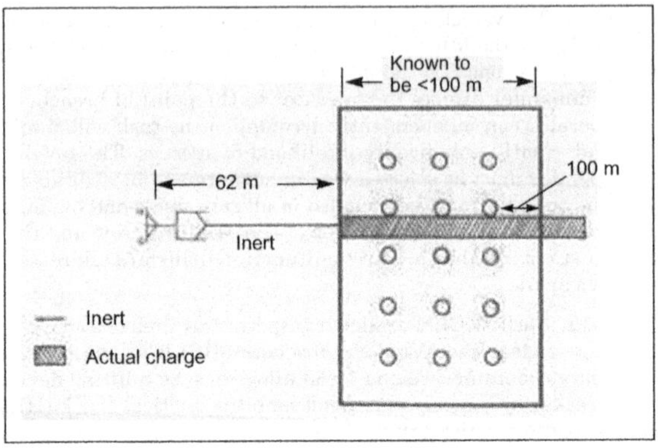

Figure C-6. MICLIC Employment in a Minefield Less Than 100 Meters Deep

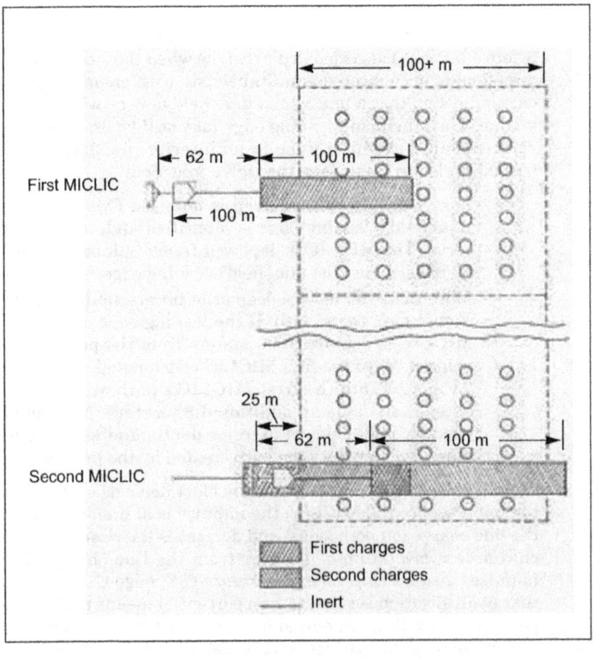

Figure C-7. MICLIC Employment in a Minefield of Uncertain Depth or Greater Than 100 Meters

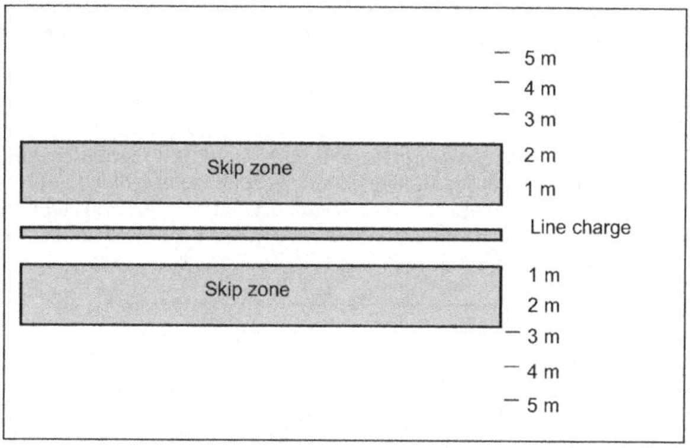

Figure C-8. MICLIC Skip Zone

Antipersonnel Obstacle Breaching System

C-20. The Antipersonnel Obstacle Breaching System (APOBS) (*Figure C-9*) is a man-portable device that is capable of quickly creating a footpath through AP mines and wire entanglements. The APOBS is normally used by combat engineers, infantry soldiers, or dismounted armored cavalry personnel. It provides a lightweight, self-contained, two-man, portable line charge that is rocket-propelled over AP obstacles from a standoff position away from the obstacle's edge.

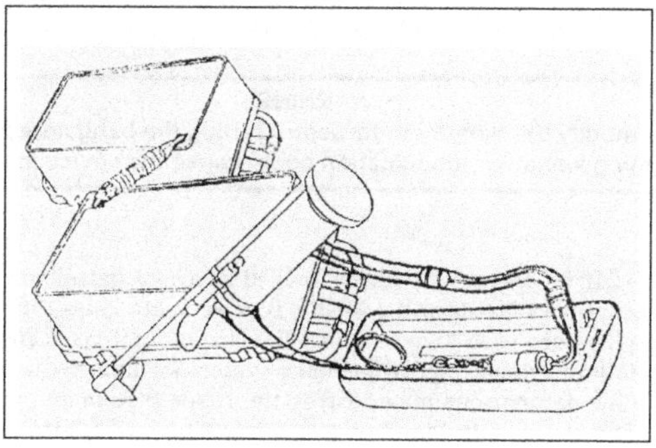

Figure C-9. APOBS

C-21. For dismounted operations, the APOBS is carried in 25-kilogram backpacks by no more than two soldiers for a maximum of 2 kilometers. One backpack assembly consists of a rocket-motor launch mechanism containing a 25-meter line-charge segment and 60 attached grenades. The other backpack assembly contains a 20-meter line-charge segment and 48 attached grenades.

The total weight of the APOBS is about 54 kilograms. It is capable of breaching a footpath about 0.6 by 45 meters and is fired from a 25-meter standoff.

M1A1/M1A2 Bangalore Torpedo

C-22. The bangalore torpedo (*Figure C-10*) is a manually emplaced, explosive-filled pipe that was designed as a wire breaching device, but it is also effective against simple pressure-activated AP mines. It is issued as a demolition kit and consists of ten 1.5-meter tubes. Each tube contains 4 kilograms of HEs and weighs 6 kilograms. The kit clears a 1- by 15-meter lane.

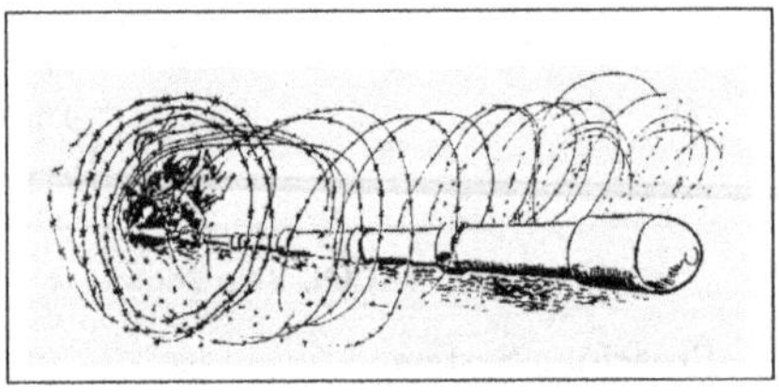

Figure C-10. Bangalore Torpedo

C-23. The bangalore torpedo is used by dismounted infantry and engineer soldiers. An individual soldier or a pair of soldiers connects the number of sections needed and pushes the torpedo through the AP minefield before priming it. A detailed recon is conducted before using the bangalore torpedo to ensure that trip wires have not been used. The bangalore torpedo generates one short impulse and is not effective against pronged, double-impulse, or pressure-resistant AP and AT mines.

WARNING

Do not modify the bangalore torpedo. Cutting the bangalore in half or performing any other modification could cause the device to explode.

MECHANICAL

C-24. MCBs and MCRs are fielded as armor battalion countermine sets that contain 12 MCBs and 4 MCRs. Blades create lanes through minefields, while rollers are used to detect minefields and proof lanes created by other means. Rollers are not a good primary system for lane reduction because multiple mine detonations may destroy the roller system and the vehicle pushing it. (The roller is designed to resist four conventional-mine [22.5 pounds HE] or three SCATMINE strikes, depending on the mine type.)

C-25. The ACE and dozer blades were not designed for reducing minefields and should only be used as a last resort. Using them is extremely time-consuming and hazardous to the crew and equipment. However, ACEs can effectively clear a lane through AP scatterable minefields because they sustain little or no damage and offer protection to the crew. When using a dozer to clear a path through a minefield, the operator is exposed to mine effects. Before clearing begins, sandbag or *up-armor* the operator's cabin and clear the lane of trip wires. When using an engineer blade to clear a path through a scatterable minefield, use the skim technique (*Figure C-11*). Start skimming 100 meters from the suspected leading edge of the minefield.

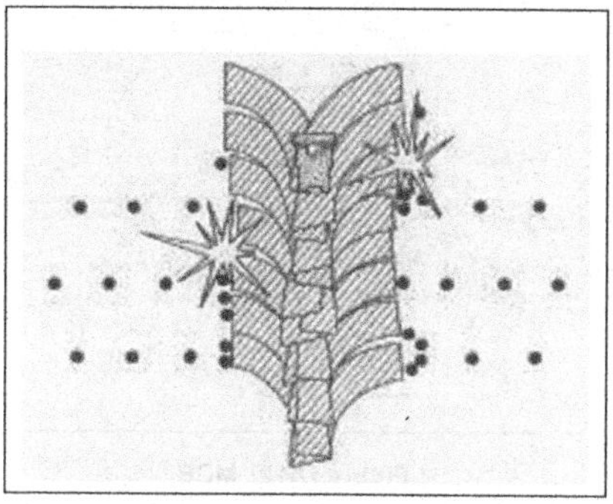

Figure C-11. Skim Technique

C-26. The ACE and dozer blades are very effective against AT ditches and gaps. Normally, during a breaching operation, dozers will not be available due to their inability to keep pace with the attacking forces. Although a single dozer or an ACE is capable of reducing an AT ditch, a technique that has been successfully used and has decreased the time to reduce an AT ditch/gap is to use two ACEs or an ACE/dozer team. When using this technique, position one ACE parallel to the AT ditch and push dirt in front of the second ACE or dozer that is oriented perpendicular to the AT ditch. Use an ACE to push dirt into the AT ditch.

Mine-Clearing Blade

C-27. The MCB (*Figure C-12*, page C-20) is used to remove land mines and create a lane in a minefield. It consists of a blade arrangement with scarifying teeth to extract mines, a moldboard to cast mines aside, and leveling skids to control the blade's depth.

C-28. The MCB lifts and pushes mines, which are surface-laid or buried up to 31 centimeters deep, to the side of track-width lanes. The blade has three depth settings—21, 25, and 31 centimeters. The blade creates a 1.5-meter cleared path in front of each track. *Figure C-13*, page C-20, shows the inside clearance distances between the tracks of common tracked vehicles in relation to the uncleared area left by the MCB.

C-29. Mines armed with AHDs, magnetic fuses, or seismic fuses may be activated when they are lifted by the blade; and they may disable the blade. Mines lifted by the blade are left in the spoil on each side of the furrowed path and remain a hazard until they are removed. Double-impulse mines that are lifted into the spoil on the side can function into the hull of the plowing vehicle.

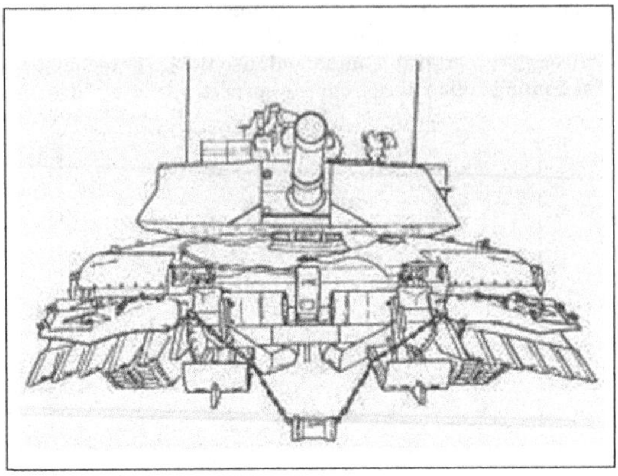

Figure C-12. MCB

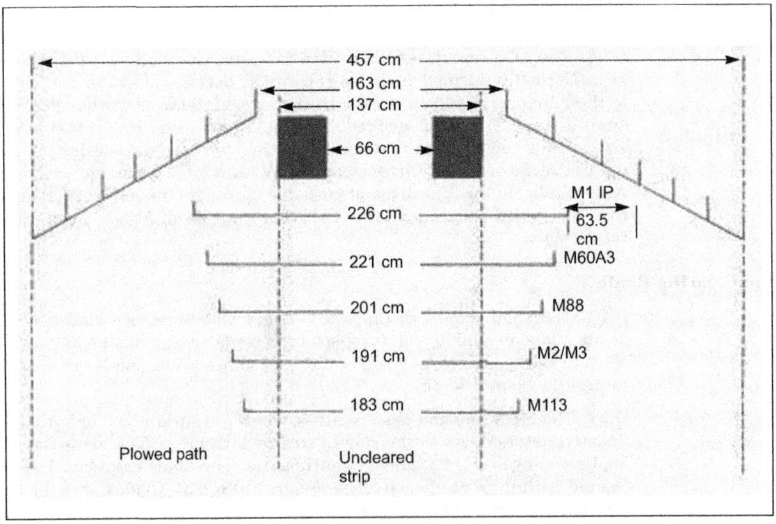

Figure C-13. MCB Width Compared to Tracked-Vehicle Widths

C-30. The skid shoe for each blade exerts adequate pressure to activate most single-impulse mines, which effectively clears a section of the centerline by explosive detonation. This action may disable the blade. Multiple-impulse pressure fuses encountered by the skid shoe are not defeated. A dog-bone assembly between the blades defeats tilt-rod mines. The improved dog-bone assembly (IDA) projects a magnetic signature and defeats tilt-rod and magnetic mines.

C-31. The MCB weighs about 3,150 kilograms and can be mounted on an M1 tank without special preparation or modification. Mounting requires lift capability and takes up to an hour, so it must be mounted well in advance of the mission. It is not easy to mount or transfer the MCB to another tank under battle-space conditions.

C-32. Once mounted, an electric motor raises and lowers the blade. When the blade is raised, it minimally effects the M1's maneuverability and speed. This will not greatly impact the use of the weapon system except when the blade is in operation. The MCB is also equipped with an emergency, quick-disconnect feature.

C-33. The M1 tank crew should maintain a speed of 8 to 10 kph while plowing, depending on soil conditions. The tank cannot maneuver but must continue in a straight path through the minefield to avoid damaging the blade. The main gun must be traversed to the side during plowing because mine detonation under the blade may cause the gun to be thrown violently into the air, damaging the tube. The area selected for the lane must be relatively flat and free of rocks or other obstacles.

C-34. The operator begins plowing about 100 meters from the estimated leading edge of the minefield. He creates a lane extending another 100 meters beyond the estimated far edge of the minefield to ensure that the lane extends through the entire minefield. Multiple vehicles using the lane will deepen the cut made by the MCB, and pressure-fused mines left in the uncleared strip will be dangerous. The uncleared strip should be cleared as soon as possible.

Mine-Clearing Roller

C-35. The MCR (*Figure C-14*, page C-22) consists of a roller assembly, a mounting kit, and a hand winch. The roller assembly weighs about 9,072 kilograms and consists of two push beams mounted to the front of the tank. The rollers are designed to defeat most single-pulse, pressure-activated AT and AP mines. The roller creates a 1.1-meter-wide cleared path in front of each track.

C-36. *Figure C-15*, page C-22, shows the inside clearance distances between the tracks of common tracked vehicles in relation to the uncleared area left by the MCR. An IDA and a chain between the rollers defeats tilt-rod mines. The IDA can be fitted to the roller. The roller is designed to withstand multiple mine explosions before damage; however, this depends on the size of the mines. Large blasts may destroy the roller or the vehicle or injure the crew.

C-37. The roller can be mounted on an M1 or M60 tank that is modified with a permanently attached mine-roller mounting kit. Mounting the roller to a tank is a cumbersome, time-consuming operation because it is very difficult

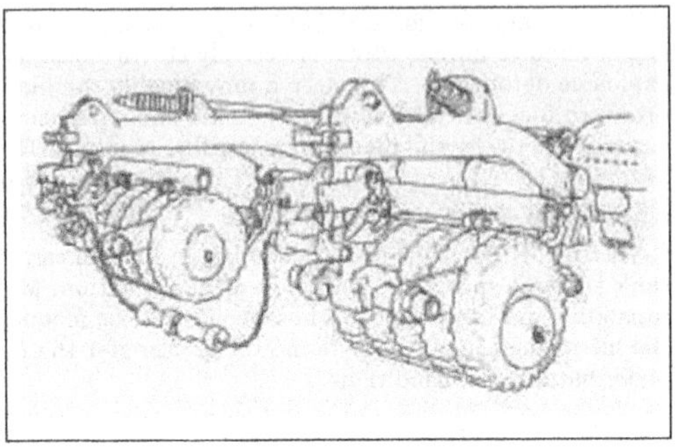

Figure C-14. MCR

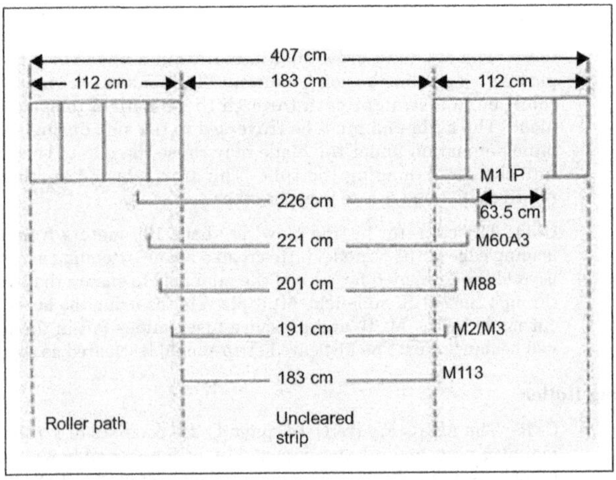

Figure C-15. MCR Width Compared to Tracked-Vehicle Width

under battle-space conditions and requires lift capability. The roller tank is limited to a speed of 5 to 15 kph. When used in a suspected minefield, the MCR must travel in a relatively straight path. Tight turns may cause the roller to deviate from the path of the track and leave the tank vulnerable to mines. Ground fluctuations, bumps, and berms may cause the roller to lift from the ground and miss mines.

C-38. Due to the roller tank's weight and width, units must use caution when crossing an AVLB. In this situation, the curbing from the bridge is removed. To prevent damage to the bridge's hydraulic line, the tank driver uses a strap to lift the IDA and the chain when crossing the bridge. The main gun must be traversed to the side when a mine encounter is possible or imminent. A mine blast can throw the roller or parts of the roller violently into the air and damage the tube. The main gun should only be fired from a temporary halt.

C-39. When the situation and the mission permit, MCRs are used with lead/security elements of the breach force to detect the minefield's leading edge. Rollers are also used to proof lanes in obstacles that are reduced by other means, such as a MICLIC or an MCB. The roller may also be used to lead a supported element traveling in a tactical formation other than a column, but it is less effective than other methods because—

- A vehicle that is not directly behind the roller may encounter mines.
- The roller may travel well into or completely through a widely spaced minefield without encountering a mine, thus giving the formation a false sense of security.
- A mine encountered by the roller may not be on the minefield's leading edge.
- The roller vehicle is extremely vulnerable since it can only use its weapon system from a temporary halt.

C-40. If rollers are to participate in a breaching operation, they must be mounted before rehearsals. Unmounted rollers that are not being used for the mission are carried in the TF formation on M916 tractor trailers. Rollers require lift capability (such as an M88), a secure location, and 30 to 60 minutes to mount on a tank that is fitted with a mounting kit.

Panther

C-41. The M60 Panther (*Figure C-16*) is one of several developmental countermine systems that US forces used during operations Joint Endeavor and Joint TF Eagle. The Panther is a remotely controlled vehicle with mine rollers, and it is used to proof lanes and assembly areas. The system consists of a turretless M60 tank, Israeli mine rollers, an antimagnetic actuating device, and a remote control unit (RCU) that is mounted in a separate vehicle. Additionally, a remote video camera allows the operator to see the road ahead.

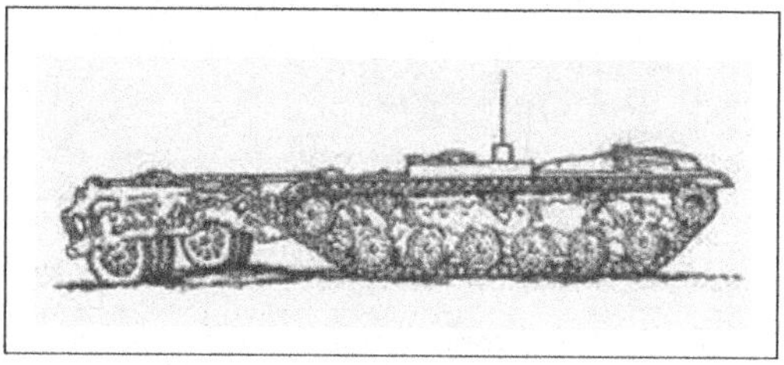

Figure C-16. Panther

C-42. During route-clearance or proofing operations, a Panther is the lead vehicle. It is followed closely by an armored control vehicle, usually an M113. The control vehicle contains the Panther operator, the RCU, and the monitor. The monitor displays the route being proofed or cleared through a camera mounted on the Panther. The Panther is controlled from the commander's cupola or troop hatch of the control vehicle. The control vehicle should be about 200 to 300 meters behind the Panther, and its hatches should be secured open. Crew members in the control vehicle should be wearing Improved Body Armor System, Individual Countermine (IBASIC) protective garments.

C-43. Mine rollers can be raised for limited travel while mounted on the Panther. If the distance is excessive, the rollers must be transported on a cargo carrier. Rollers must be adjusted before every mission to ensure that they have contact with the ground and that their weight is uniformly distributed. Conduct at least three passes to ensure proper coverage and overlap of rollers. Passes should have a minimum of 30 centimeters overlap. Inside roller distances are the same as an MCR.

M1 Grizzly

C-44. The Grizzly (*Figure C-17*) provides the capability for breaching obstacles comprised of mines, wire, posts, rubble, and tank ditches to create a lane for other vehicles to follow. The Grizzly's primary features are a full-width, 4.2-meter MCB and a power arm. The power arm has a reach of 9 meters and a bucket capacity of 1.2 cubic meters. Its primary missions are to reduce berms and fill AT ditches.

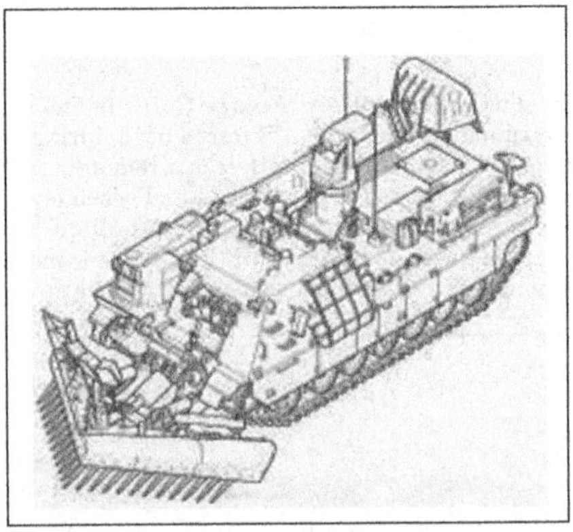

Figure C-17. Grizzly

C-45. The Grizzly lifts and pushes mines, which are surface-laid or buried up to 31 centimeters deep, to the side of full-width lanes. The blade has multiple depth settings, depending on the mission, and it creates a 4.2-meter-wide cleared path. When plowing, the Grizzly is restricted to less than 10 kph, depending on soil conditions. The operator begins plowing about 100 meters from the estimated leading edge of the minefield. He creates a lane extending another 100 meters beyond the estimated far edge of the minefield to ensure that the lane extends through the entire minefield.

C-46. The Grizzly has integrated digital features to enhance battlefield awareness. Some of the digital features are thermal and video cameras, ground-speed sensors, terrain-mapping sensors, and an integrated commander's control station.

Mine-Clearing/Armor-Protection Kit

C-47. The mine-clearing/armor-protection (MCAP) kit consists of two parts—the mine-clearing rake and the armor protection.

- **Mine-clearing rake.** The rake uses a tine that is mounted on a diagonal beam. The rake assembly is designed to sift through the soil, lift out mines, and windrow buried and surface-laid mines to the right of the vehicle. The system clears a 30-centimeter-deep path through a minefield. The rake has a skid shoe that acts as a depth control guide for the operator.
- **Armor protection.** The armor protects the crew against mine blast, small-arms fire, and artillery fire. Protection is also provided for the engine, the fuel tank, and exposed hydraulic cylinders and lines. Ballistic glass blocks are provided at each vision port to permit unrestricted view and operation of the vehicle and the equipment.

The MCAP is mounted on a D7 dozer to perform minefield reduction and lane widening. Proofing the lane must be conducted after the dozer has cleared the lane because some AP mines may still be left in the lane.

Armored Vehicle-Launched Bridge

C-48. The AVLB (*Figure C-18*) is carried on a launcher (a modified M48A5 or M60A1 chassis) and weighs 15 tons. A buttoned-up, two-man crew can launch the AVLB in 2 to 5 minutes. The AVLB can be retrieved from either end; however, one man is exposed to fire while guiding and connecting it. The AVLB is capable of carrying military load class (MLC) 60 (70 with improved bridge) tracked loads across a 17.4-meter gap with unprepared abutments or an 18.3-meter gap with prepared abutments.

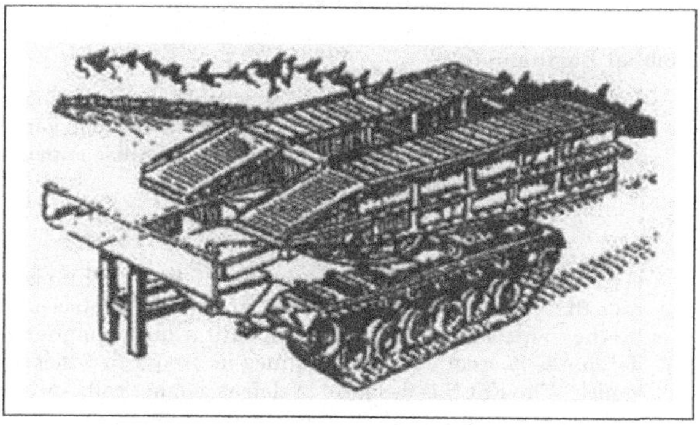

Figure C-18. AVLB

C-49. The AVLB has the following limitations:

- The scissor launch requires 10 meters of overhead clearance.
- The maximum launch slope is—
  - 28 percent uphill.
  - 19 percent downhill.
  - 11 percent on a side slope.

Wolverine

C-50. The AVLB will eventually be replaced by the Wolverine (*Figure C-19*). The bridge is carried on a launcher (a modified M1 tank chassis) and weighs 12.5 tons. It can be launched in less than 5 minutes by a buttoned-up, two-man crew and can be retrieved from either end of the bridge in less than 10 minutes. The Wolverine can cross an MLC 70 tracked vehicle over 24-meter gaps. The launcher has a maximum speed of 83 kph.

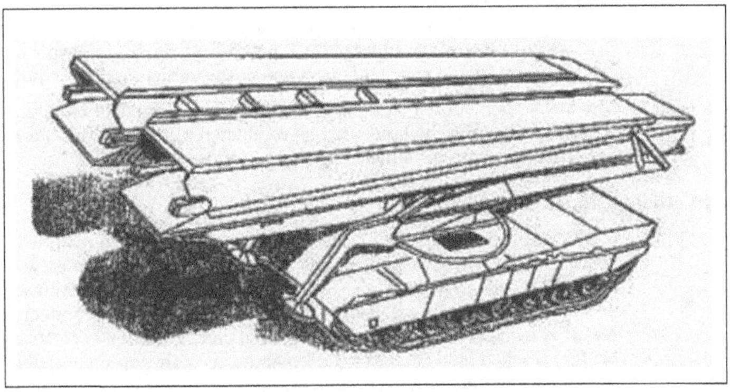

Figure C-19. Wolverine

M9 Armored Combat Earthmover

C-51. The M9 ACE (*Figure C-20*) is capable of performing mobility tasks in support of light and heavy forces. These tasks include gaps and AT ditches. Although an ACE can be used to reduce a scatterable minefield, it should only be used as a last resort.

ELECTRONIC

C-52. The Field-Expedient Countermine System (FECS) is a series of copper coils that fit over the front of tracked and wheeled vehicles. Power is supplied by the vehicle's battery. The coils emit a large magnetic signature that detonates magnetically fused mines located 2 to 5 meters in front of the vehicle. The FECS is designed to defeat magnetically influenced mines only and must be used with other countermine systems.

Figure C-20. M9 ACE

MANUAL

C-53. While mechanical and explosive reduction procedures are normally preferred, a unit may have to use manual procedures for the following reasons:

- Explosive and mechanical mobility assets are unavailable (not on hand or nonmission capable).
- Explosive and mechanical mobility assets are ineffective against the type of obstacle or because of terrain limitations (see *Table C-1*, page C-5).
- Stealth is required.

C-54. Manual procedures involve dismounted soldiers (usually engineers) using simple explosives or equipment to create a lane through an obstacle or to clear an obstacle. These procedures expose the soldier and may be manpower- and time-intensive. However, this method works in most situations and under most conditions. Certain types of terrain, weather, and sophisticated fuses can severely degrade the effectiveness of rollers, plows, and line charges.

Techniques

C-55. As discussed, manual procedures are effective against all obstacles under all conditions. Most types of obstacles do not cause casualties directly, but minefields have the potential to cause direct casualties. The enemy possesses a significant mechanical, mine-burying capability. It has the capacity and the propensity for the labor-intensive effort required to bury mines by hand; however, the enemy often lays mines on the surface. Buried mines are usually found in a highly prepared defense. When training for the reduction of surface-laid and buried minefields, always assume the presence of AHDs and trip wires until proven otherwise. Manual reduction of surface-laid and buried minefields is discussed below:

- **Surface-laid minefield.** Engineers first use grapnel hooks from covered positions to check for trip wires in the lane. The limited range of the tossed hook requires the procedure to be repeated through the estimated depth of the obstacle. A demolition team then moves through the lane. The team places a line main down the center of the lane, ties the line from the explosive into the line main, and places blocks of explosive next to surface-laid mines. After the mines are detonated, the team makes a visual check to ensure that all mines were cleared before directing a proofing roller and other traffic through the lane. Members of the demolition team are assigned special tasks, such as grappler, detonating-cord man, and demolitions man. All members should be cross-trained on all procedures. Demolitions are prepared for use before arriving at the point of breach. An engineer platoon uses squads in series through the minefield to clear a lane for a company team. The platoon must rehearse reduction procedures until execution is flawless, quick, and technically safe. During reduction, the engineer platoon will be exposed in the lane for 5 minutes or more depending on the mission, the minefield depth, and the platoon's level of training.

- **Buried minefield.** Manually reducing a buried minefield is extremely difficult to perform as part of a breaching operation. If mine burrows are not easily seen, mine detectors and probes must be used to locate mines. Mines are then destroyed by hand-emplaced charges. As an alternative, mines can be removed by using a grappling hook and, if necessary, a tripod (*Figure C-21*). Using a tripod provides vertical lift on a mine, making it easier to pull the mine out of the hole.

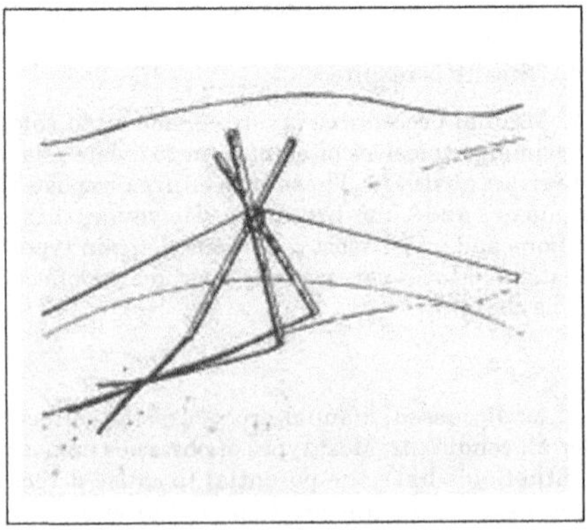

Figure C-21. Tripod

C-56. The platoon leader organizes soldiers into teams with distinct, rehearsed missions including grappling, detecting, marking, probing, and emplacing demolitions and detonating cord. The platoon is exposed in the obstacle for long periods of time.

Grappling Hook

C-57. The grappling hook (grapnel) is a multipurpose tool that is used for manual obstacle reduction. Soldiers use it to detonate mines from a standoff position by activating trip wires and AHDs. After the grapnel is used to clear trip wires in a lane, dismounted engineers can move through the minefield, visually locate surface-laid mines, and prepare mines for demolition. In buried minefields, soldiers grapple, and then enter the minefield with mine detectors and probes. There are two types of grapnel hooks—hand-thrown and weapon-launched.

- **Hand-thrown.** A 60+-meter light rope is attached to the grapnel for hand throwing. The throwing range is usually no more than 25 meters. The excess rope is used for the standoff distance when the thrower begins grappling. The thrower tosses the grapnel and seeks cover before the grapnel and rope touch the ground in case their impact detonates a mine. He then moves backward, reaches the end of the excess rope, takes cover, and begins grappling. Once the grapnel is recovered, the thrower moves forward to the original position, tosses the grapnel, and repeats the procedure at least twice. He then moves to the end of the grappled area and repeats this sequence through the depth of the minefield.
- **Weapon-launched.** A 150-meter lightweight rope is attached to a lightweight grapnel that is designed to be fired from an M16A1 or M16A2 rifle using an M855 cartridge. The grapnel is pushed onto the rifle muzzle with the opening of the retrieval-rope bag oriented toward the minefield. The firer is located 25 meters from the minefield's leading edge and aims the rifle muzzle at a 30- to 40-degree angle for maximum range. Once fired, the grapnel travels 75 to 100 meters from the firer's position. After the weapon-launched grapnel hook (WLGH) has been fired, the firer secures the rope, moves 60 meters from the minefield, moves into a prone position, and begins to grapple. The WLGH can be used only once to clear a minefield, but it can be reused up to 20 times for training because blanks are used to fire it.

C-58. Multiple grapplers can clear a lane of trip wires quickly and thoroughly, but they must time their efforts and follow procedures simultaneously. A hit on a trip wire or a pressure fuse can destroy the hook and the cord, so engineers should carry extras.

Appendix D

Breaching Scenario and Special Planning Considerations

This appendix provides a scenario showing how the breaching tenets may be applied when conducting a breaching operation in open terrain, such as a desert. This example is not an attempt to give a single, doctrinally correct solution but simply to provide a method that a unit may use to task-organize and approach this very difficult mission. Also discussed in this appendix are considerations that may be needed when conducting a breaching operation in restricted terrain (a defile, urban terrain, or the jungle). The following paragraphs are an example of a worst-case scenario in which a TF is given the mission to conduct a breaching operation against a well-equipped, well-prepared enemy defense.

SAMPLE BREACHING OPERATION

D-1. TF 1-12 is a mechanized infantry battalion that is task-organized with two tank companies from TF 3-71 (armored battalion). It just received a mission to secure OBJ SPARROW as part of a brigade attack to destroy a motorized infantry battalion (MIBN) (see *Figure D-1,* page D-2).

D-2. TF 1-12 received intelligence products, including the brigade SITEMP and a series of WOs, before issuing the actual OPORD. The TF staff, in particular the S2 and the engineer, revised and refined the brigade SITEMP based on their analyses of the terrain and the enemy (see *Figure D-2,* page D-3).

D-3. Upon receipt of the brigade OPORD, TF 1-12 learns that—

- The brigade has task-organized it with an engineer recon team to augment its recon capability and its habitually associated engineer company. An additional engineer line platoon has been attached to the engineer company, giving the TF a total of three engineer line platoons.
- The brigade has given TF 1-11 the mission to fix enemy forces north of OBJ SPARROW to prevent them from repositioning.
- The brigade plans to disrupt the enemy's counterattack force by using area-denial artillery munition (ADAM)/remote antiarmor mine (RAAM) SCATMINEs along the high-speed route to OBJ SPARROW. The counterattack force has been located by intelligence sources.
- The brigade plans to have the engineer battalion (-) follow TF 1-12's assault force to assume control of the lanes. The engineer battalion will create additional lanes, if necessary, and upgrade/maintain the lanes and marking pattern to facilitate the passage of TF 3-71.

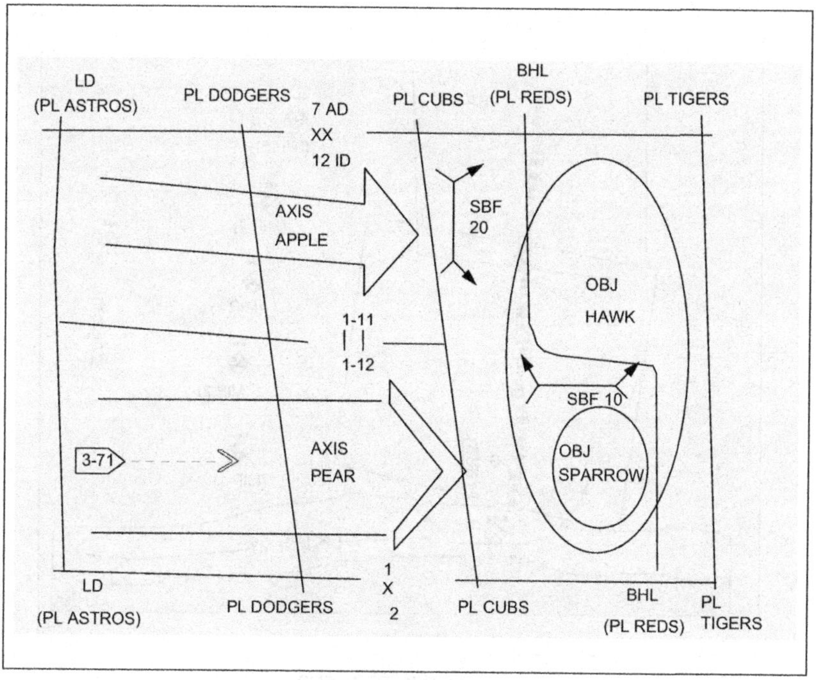

Figure D-1. Brigade Maneuver Graphics

- The brigade has tasked it to seize OBJ SPARROW and occupy SBF 10 with at least two platoons to fix enemy forces north of OBJ SPARROW.
- The brigade has tasked it to create two lanes in the obstacle system to support the passage of TF 3-71 through the created lanes.

D-4. Based on the commander's guidance, the TF staff developed three COAs. The third COA was chosen based on war gaming. It required a company team (assault force), consisting of one tank platoon (the platoon has two tank plows mounted) and two mechanized infantry platoons, to seize OBJ SPARROW. The staff decided that two lanes were necessary in the obstacle system, vicinity grid AA123456, based on the specified task given by the brigade. Because of the lane requirement and the obstacle type expected to be encountered, the staff decided that the breach force must consist of at least two tanks with MCBs, two tanks with MCRs, and two MICLICs. The breach force must also have two engineer platoons to create, proof, and mark the lanes and to provide redundancy of assets. The staff considered the threat to the breach force to be—

- The infantry platoon and AT weapons on the southern flank.

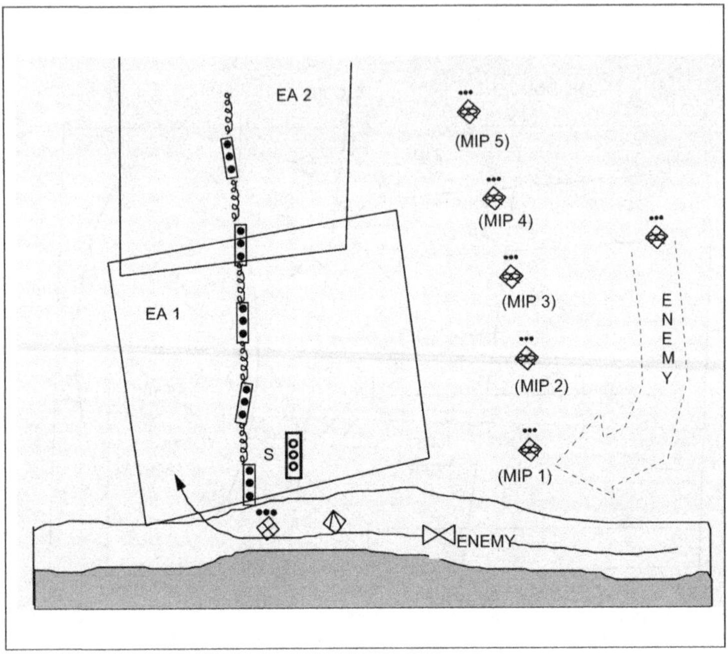

Figure D-2. TF SITEMP

- The southern motorized infantry platoon (MIP), which the support force may not be able to suppress because its forces are masking the breach force.
- The MIP-size counterattack force.

D-5. In this COA, the staff task-organized one tank platoon and one mechanized infantry platoon to the breach force (in addition to the MCBs and MCRs) to provide local security and redundant suppression. The staff determined from its terrain and enemy analyses that the support force's mission is to suppress the four southern MIPs to facilitate reduction operations. To accomplish this, the staff task-organizes two company teams to the support force. Each company team in the support force has one tank plow to reduce any obstacles it encounters en route to the SBFs (see *Figure D-3,* page D-4).

D-6. The staff further refines the COA by developing an initial plan to achieve the following breaching fundamentals, to include assigning tasks to the various subordinate forces (see *Figure D-4,* page D-5):

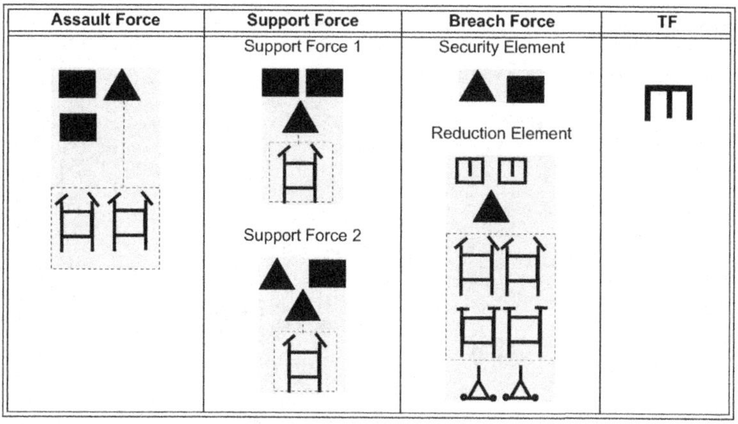

Figure D-3. Initial Breaching Organization

- **Suppress.**
  - Support force 1 will occupy SBF 1 and initially orient its fires on MIPs 1 and 2. When the breach force is committed, it will shift its fires to orient on MIPs 2 and 3. Fires will be lifted on the order of the assault-force commander.
  - Support force 2 will occupy SBF 2 and orient its fires on MIPs 3 and 4. Fires will be lifted on the order of the assault-force commander.
  - The breach-force security element will occupy SBF 3 and orient fires on MIP 1. The security element is responsible for local security at the point of breach.
  - The fire-support element (FSE) will ensure that CFZ 1 is activated as the breach force moves to the point of breach.
  - The support force will call for and adjust artillery targets AG7001 through AG7003 when the lead elements of the support force cross PL CUBS.
- **Obscure.**
  - The support force will call for and adjust smoke target AG1210 when the lead elements of the support force cross PL CUBS. The support force will adjust this target only if the smoke is ineffective.
  - The smoke platoon (task-organized to the TF for this mission) will employ its generators to the north of support force 2. Smoke operations begin when support force 2 crosses PL CUBS.
- **Secure.** The breach force (security element) will occupy SBF 4 and orient fires on the infantry platoon and AT weapons on the south wall.

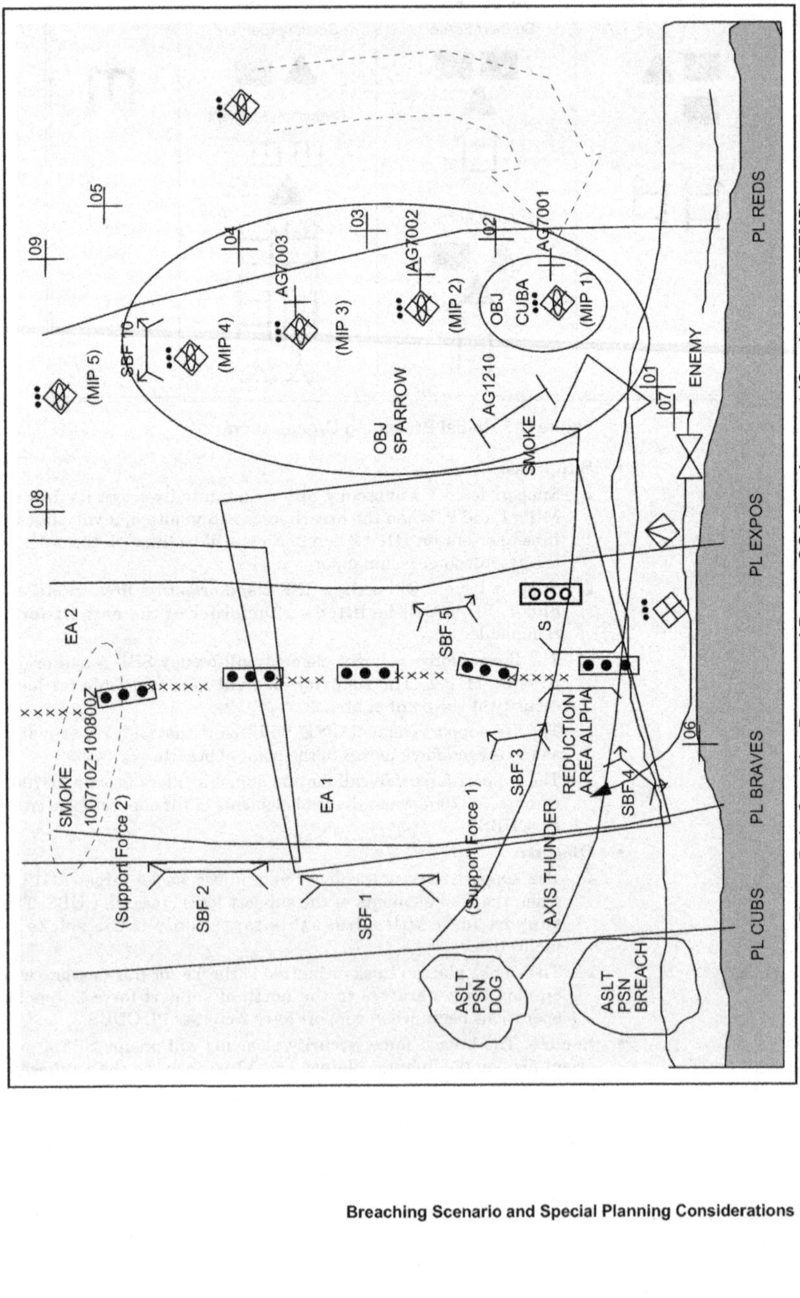

Figure D-4. Graphics Developed During COA Development (Overlaid on SITEMP)

- **Reduce.**
  - The breach-force reduction element will move from assault position (ASLT PSN) BREACH when the commitment criteria in reduction area ALPHA has been met. In this scenario, it will create, proof, and mark (initial lane-marking pattern) two lanes in the obstacle system, at least 800 meters apart, within reduction area ALPHA. This considers the enemy's use of artillery-delivered SCATMINEs to close the lanes.
  - The breach-force security element will occupy SBF 5 with one platoon orienting fires on MIPs 2 and 3 to provide enemy suppression when one lane is created in the tactical obstacles.
- **Assault.**
  - The assault force will attack along axis THUNDER to initially seize OBJ CUBA (point of penetration) and then the rest of OBJ SPARROW.
  - The assault force will occupy SBF 10 with at least two platoons oriented on MIP 5.

D-7. During COA analysis, modifications are made to the original COA. This includes task-organizing an engineer platoon to the assault force to help reduce protective obstacles and mark lanes for the assault and follow-on forces (see *Figure D-5*).

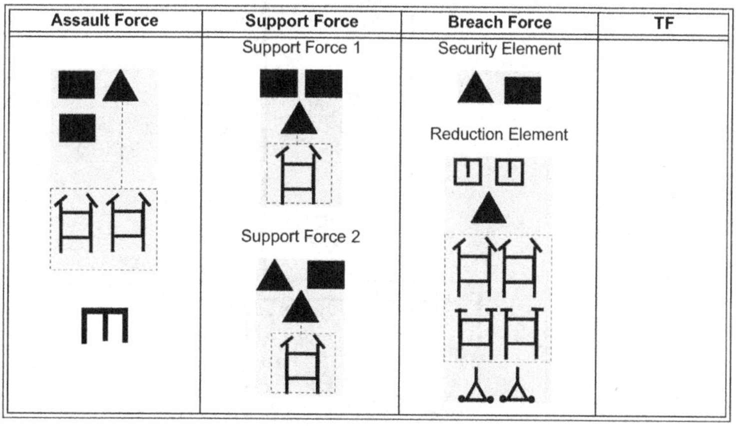

Figure D-5. Final Task Organization

D-8. Additionally, the staff develops a decision point for committing the breach force and a plan to collect the necessary information to support the decision point (see *Table D-1*). *Figure D-6* shows the location of each NAI.

Table D-1. Plan to Support the Decision Point

Decision Criteria	NAI	Responsibility
MIP in OBJ CUBA must be destroyed.	NAI 300	TF scouts
Enemy south of TRP 5 must be suppressed.	NAI 301	Support force commander
Point of breach must be effectively obscured.	NAI 302	Support force commander
Support force must have a minimum of 16 M1s/M2s available.		Support force commander
Breach force must have a minimum of one engineer platoon, 1 MICLIC/ MCB, and 6 M1s/M2s.		Breach force commander
CFZ 1 must be activated.		TOC (FSE)

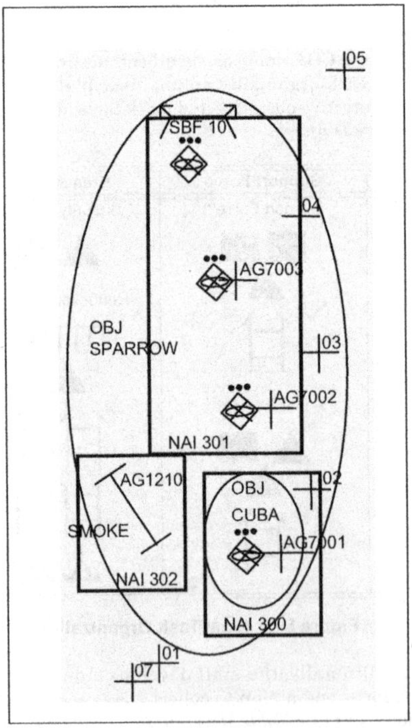

Figure D-6. NAIs to Support the Decision Point

D-9. The commander specifies the following C^2 structure of the mission based on the scheme of maneuver in the COA he approved:

- The assault force will be comprised of Team A/3-71 (mechanized heavy).
- The breach force will be comprised of Team B/3-71 (tank heavy) and A/ 55th Engineer Battalion. The Team B commander will be the breach-force and security-element commander. The engineer commander will be the reduction-element commander.
- The support force will be comprised of Team C/1-12 (mechanized heavy) and Team D/1-12 (tank heavy). The support-force commander will be the TF S3.

D-10. Preparation for the breaching operation involves numerous rehearsals, beginning at the lowest levels and culminating into a full-dress, combined-arms rehearsal. S2 reports that the TF received the confirmed location of the tactical obstacles in the vicinity of the point of breach. The report indicated that the minefields are comprised of three rows of buried SB-MV mines (Italian AT mines), with a uniform 4.5 meters spacing between mines and 70 meters between rows. One row of concertina wire is positioned 40 meters to the rear of each tactical minefield. Additionally, the enemy has emplaced protective minefields, which are comprised of AP mines only, 400 meters in front of each platoon position. The enemy has tanks and infantry fighting vehicles (IFVs) overwatching tactical and protective obstacles. Reports indicate that the enemy has not prepared dismounted positions.

NOTE: The assault-force commander decides to conduct a mounted assault based on information from the intelligence report and the terrain analysis he conducted. The terrain analysis indicates that tank plows will effectively reduce the enemy's protective obstacles.

D-11. This information is quickly relayed to subordinate units and all rehearsal sites reflect these obstacles. The commander pays particular attention to the status and location of reduction assets throughout preparations for the attack. This results in an MCB being removed from an inoperable M1 at the unit maintenance collection point (UMCP) to an operable M1 in the assault force.

D-12. The TF crosses the LD at 95 percent strength (all MCBs, MICLICs, and MCRs are operational). Before crossing PL DODGERS, the TF scouts report that three of the four vehicles in NAI 300 have been destroyed by Copperhead munitions. The commander reminds the S3 that destroying the remaining vehicles is his first priority.

D-13. As the support force crosses PL CUBS, the TF S3 reports that he has lost two M2s to indirect fires and one M1 to maintenance and is beginning to occupy SBFs 1 and 2. Artillery units begin suppressive fires and begin to build obscuration smoke. As the support force begins to suppress the enemy with direct and indirect fires, the breach and assault forces occupy ASLT PSNs BREACH and DOG, respectively, and conduct final preparations as necessary. The tactical operations center (TOC) informs the commander that CFZ 1 will be activated when the breach force is committed. The breach-force commander reports that all reduction assets are operable, and the TF scouts report that the final vehicle in OBJ CUBA has been destroyed. The TF S3 reports that obscuration and suppression have been achieved and that combat power is 6 M1s and 13 M2s.

D-14. Having met the commitment criteria, the commander orders the breach force to AA123456 to begin obstacle reduction. The breach-force commander initially moves his security element to SBFs 3 and 4. The reduction-element commander also moves forward with his platoon leaders to recon the obstacle and confirm/refine the plan to reduce the obstacle. Once the security element is set in SBFs 3 and 4, the reduction element moves forward to reduce the obstacle. When the first lane is created and proofed, the breach-force commander moves a portion of the security element from SBF 3 to SBF 5. The assault-force commander begins movement from the assault position to the far-recognition marker that has been reported by the reduction-element commander. Upon reaching the far-recognition marker, the assault-force commander can observe the final-approach marker and the marked lane.

D-15. The assault force moves through lanes in the tactical obstacles and sets its platoons in the SBFs. It immediately begins to create two lanes in the protective obstacles with the MCBs. When the lanes are complete, the assault-force commander maneuvers the two tank platoons to SBF 10, while the mechanized company team continues to suppress the remaining enemy vehicles. The engineer platoon remains at protective obstacles to mark lanes in preparation for passing follow-on forces.

D-16. The engineer battalion S3 is immediately behind the assault force. He takes control of the reduction element that has created, proofed, and marked two lanes in the tactical obstacles by this time. The S3 moves a portion of the reduction assets away from the lanes to counter any SCATMINEs that the enemy may use to close the lanes. The S3 prepares to assist with TF 3-71's passage through the lanes. When the assault force reports that it has created a safe route to PL REDS, the brigade commander orders TF 3-71 to pass through TF 1-12 and continue the attack.

CONSIDERATIONS FOR RESTRICTED TERRAIN

D-17. The breaching tenets discussed in *Chapter 1* apply to breaching operations in restricted terrain; however, the application of these tenets is different from a breaching operation conducted in open terrain. The following considerations should be made when breaching in restricted terrain:

- OBSTINTEL collection is critical when conducting breaching operations in restricted terrain. It is essential that the breach force knows what types of obstacles are in place and knows the exact composition of any complex obstacle. With this information, the breach-force commander can ensure that the reduction assets necessary to reduce the obstacle are available and placed in the right order in the movement formation. The breach force's ability to move reduction assets within the formation may be extremely limited due to the terrain's constricting nature.
- Recon assets tasked to obtain OBSTINTEL should be prepared to recon different types of obstacles than would normally be found in open terrain. For example, in urban or mountainous terrain, the

enemy may use rubble obstacles, such as the *rock drop*. Additionally, recon assets are much more likely to encounter field-expedient or booby-trapped obstacles when conducting operations in restricted terrain. Recon assets used to collect OBSTINTEL must be able to recommend reduction methods to be used by the breach force when reducing the obstacle.

D-18. The proper application of the following breaching fundamentals is essential to the success of breaching operations conducted in restricted terrain. As with other breaching tenets, breaching fundamentals are normally applied differently in restricted terrain than in open terrain.

- **Suppress**. The need to prevent or degrade enemy fires and observation of friendly forces remains unchanged when breaching in restricted terrain. However, unlike a breaching operation conducted in open terrain in which the support force can suppress the enemy with long-range fires, massed indirect fires, and smoke, suppression will normally be achieved through close-in fighting. Forces tasked with suppressing the enemy may find it necessary to seize terrain at or on the far side of the obstacle to achieve their mission. In all probability, the fighting to achieve enemy suppression will be decentralized. Therefore, the ability to use indirect fires will be restricted due to the close proximity of friendly and enemy forces.
- **Obscure**. When trying to obscure the enemy with indirect fires, the same problems exist as when trying to suppress the enemy. Using mortars (because of their ability to fire high-level trajectory), smoke pots, and smoke grenades may be more effective. The potential for degrading friendly-force operations when using obscurants is greater because of the close-in nature of the fighting that occurs in restricted terrain.
- **Secure**. The responsibility to secure the reduction area may normally be given to the support force when conducting breaching operations in restricted terrain. Because the support force may find it necessary to seize terrain on the far side of the obstacle to suppress the enemy, securing the reduction area may be accomplished at the same time. When this situation occurs, the amount of maneuver forces within the breach force may be less than what is normally necessary when breaching in open terrain.
- **Reduce**. The ability to create more than one lane and to move reduction assets around each other may be severely restricted when breaching in restricted terrain. For this reason, accurate OBSTINTEL and a well-thought-out, well-rehearsed reduction plan are critical.
- **Assault**. As discussed earlier, the number of lanes available to the assault force may be less than what is actually needed. For example, a battalion-size assault force may be restricted to using a single lane through an obstacle due to the restricted terrain. For this reason, the assault-force commander must ensure that the sequencing of forces through the lane is appropriate to achieve the mission.
- **Mass**. Although mass remains a critical tenet for breaching operations in restricted terrain, the ability to achieve mass may be difficult and time-consuming. The assault force will attempt to

achieve a 3:1 combat-power ratio. However, due to the restricted nature of the terrain and the limited number of lanes that may be created, the assault force may take longer to get that much combat power to the far side of the obstacle. Units must consider ways to minimize the risk to the assault force as it builds its combat power on the far side of the obstacle.

- **Synchronize**. The reverse-planning process described in *Chapter 1* remains the key to successful breaching operations in restricted terrain. However, effective C^2, particularly the controlling of personnel and equipment around the point of breach, is critical for successful operations. The terrain's restricted nature magnifies mistakes made in bringing the proper reduction assets forward to the obstacle or in sequencing forces through created lanes. These mistakes result in increased time for the enemy to react to the breaching operation through repositioning and other means. For this reason, a well-rehearsed force is critical. All subordinate units must understand their mission, the overall timing of the operation, and how their actions relate to missions of other subordinate units.

CONSIDERATIONS FOR MILITARY OPERATIONS IN URBANIZED TERRAIN

D-19. Military operations in urbanized terrain (MOUT) is defined as all military actions that are planned and conducted on a topographical complex and its adjacent natural terrain where man-made construction is the dominant feature. Urbanized terrain affects the tactical options available to the commander and requires a thorough knowledge of unique terrain characteristics, detailed planning down to the smallest unit level, and sound leadership at all levels. The complexities of the urban environment, such as line-of-sight (LOS) restrictions, inherent fortifications, limited intelligence, densely constructed areas, and the presence of noncombatants, restricts current military technology. US forces do not possess the overwhelming technology advantages in an urban environment as in other environments.

D-20. A three-dimensional view of the terrain is essential to gain a complete understanding of observation and AAs. The close fight (inside the building) is difficult and decentralized, with control of the movement from room to room most often in the hands of squad and platoon leaders. Success in the movement from building to building is a combined-arms fight, which requires the active involvement of a battalion command post. Seeing the terrain in a three-dimensional view will identify opportunities for employment of all the battalion's combat power and enable the battalion to set the conditions for successful platoon and company assaults.

PLANNING

D-21. Close-quarters clearing techniques are designed to be executed by standard four-man breach and clearing teams. Therefore, the size of the breach organization will be significantly smaller than those involved when fighting occurs in open terrain. The reduction element must work with the breach leader to formulate the assault plan and to ensure that all planning considerations are completed.

D-22. The reduction element is responsible for the following tasks during the planning phase:

- Accessing all intelligence data on potential breach points.
- Helping the command element select the best method of entry (mechanical, shotgun ballistic breach, or explosive). (See *Figures D-7 through D-9*, and *Figure D-10*, page D-14)
- Assembling and preparing tools and equipment and constructing explosive entry charges and diversionary charges if required.
- Briefing team members.
- Rehearsing.
- Conducting a back brief of team members.
- Conducting a precombat inspection (PCI).

D-23. The commander, breacher, and other essential personnel assemble and analyze all available information, data, and intelligence pertaining to the intended target. This involves examining the design construction and material makeup of the structure to determine the best primary and alternate points of breach.

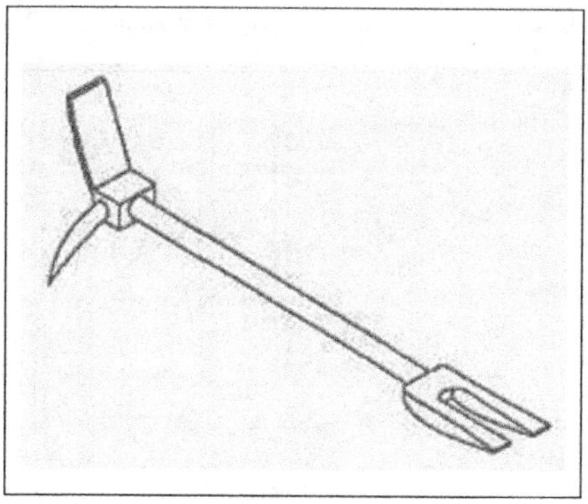

Figure D-7. Hooligan Tool

METHOD OF ENTRY

D-24. Once the point of breach has been identified, the reduction element leader continues to assemble and analyze all available information, data, and intelligence pertaining to the target structure. This involves a detailed examination of each point of breach's design, construction, and material makeup to determine the best primary, secondary, and tertiary breaching method. Again, tactics to be employed may dictate or rule out specific entry points.

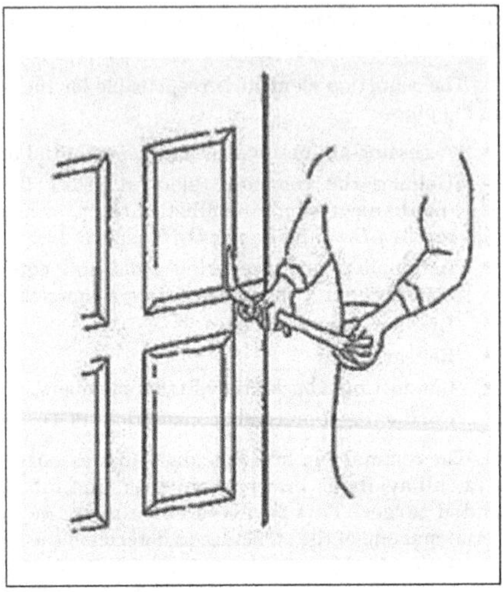

Figure D-8. Using a Hooligan Tool Against an Inward-Opening Door

Figure D-9. Attacking a Dead Bolt with a Sledgehammer

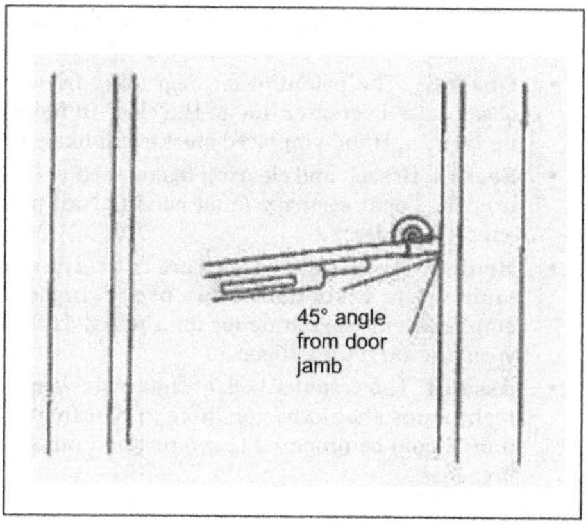

Figure D-10. Using a Shotgun Against a Door Lock

OBSTACLE INTELLIGENCE

D-25. Collecting the following OBSTINTEL is critical during MOUT:

- What type of mission is it?
- Are drawings or blueprints of the structure available?
- Are photographs of the point of breach and adjacent structures available?
- Can R&S personnel give a detailed description of the point of breach?
- Can counterintelligence personnel provide any information?
- Are there any human intelligence sources (agents, friendly forces, evacuees, prisoners, tourists, workers, businessmen, travelers) that can provide information relating to the point of breach?
- What type of shielding is available at potential breach points (natural or man-made)?
- How close will the assault force be positioned to the breach points?
- Do drawings of the target (if available) include gas, power lines, and other similar hazards?
- Is there a history of booby-trap use?
- Will the weather affect charge adhesion methods?

BREACHING FUNDAMENTALS

D-26. The proper application of the following breaching fundamentals is essential to ensure success in breaching conducted during MOUT. As with other breaching tenets, the breaching fundamentals are normally applied differently in urbanized terrain than in open terrain.

- **Suppression**. Maneuvering to achieve suppression of the enemy will be decentralized. Therefore, the ability to use indirect fires will be restricted due to the close proximity of friendly and enemy forces and civilians.

- **Obscure**. The potential for degrading friendly-force operations with obscurants is greater due to the close-in fighting that occurs in built-up terrain. Hand-emplaced smoke will likely be the most effective.
- **Secure**. Breach and clearing teams secure a foothold near the point of breach. Local security must consider all possible locations for the enemy to hide.
- **Reduce**. Redundant systems are critical during MOUT. The reduction element is responsible for preparing equipment/charges for emplacement, searching for improvised explosive devices (IEDs), and verifying target intelligence.
- **Assault**. The actual attack of the objective using close-quarters battle techniques should be conducted according to the assault plan. The unit should be prepared to conduct additional planned and unplanned breaches.

Appendix E

Route Clearance

Route clearance is a combined-arms operation typically executed by an infantry company or a cavalry troop. Units must clear lines of communication (LOC) of obstacles and enemy activity that disrupt battle-space circulation. Units must conduct route clearance to ensure that LOC enable safe passage of combat, combat-support (CS), and CSS organizations. Clearance operations are normally conducted in a low-threat environment.

METHODS AND LEVELS

E-1. The information gathered from the IPB and the recon effort determines the method and type of route clearance to conduct. The determination is based on the situation, the time available, the threat level, and available assets.

E-2. During stability and/or support actions, it is recommended that former-warring-faction (FWF), engineer-equivalent clearance teams precede US clearance teams within the FWF's AO. Do not assume that FWF clearance teams will be thorough in their clearance operation. Treat the route as unsafe until US or allied force clearance teams have proofed the route to confirm that it is cleared.

CLEARANCE METHODS

E-3. There are three methods of route clearance—linear, combat, and combat route. The method employed depends on the situation, the time available, and the clearance assets available. The maneuver force should always establish static security positions at critical locations following the completion of route clearance.

- **Linear**. In linear clearance (*Figure E-1,* page E-2), the clearance begins at checkpoint 1 and is completed at checkpoint 2. This method provides the best assurance of route clearance. Although this is an effective method, it is not the most secure method in a high-threat environment. It is also time-intensive and constrains the maneuver commander's flexibility.
- **Combat**. Whereas linear clearance focuses on a specific route, combat clearance (*Figure E-2,* page E-2) focuses on specific points along a route. As mentioned previously, the IPB can identify likely areas for obstacle and ambush locations. These areas become NAIs or objectives for combat clearance missions. The combat clearance method divides a route into sections according to the number of suspected high-threat areas. Once the unit clears these areas, the route is considered clear. Combat forces can patrol the route from these objectives to ensure

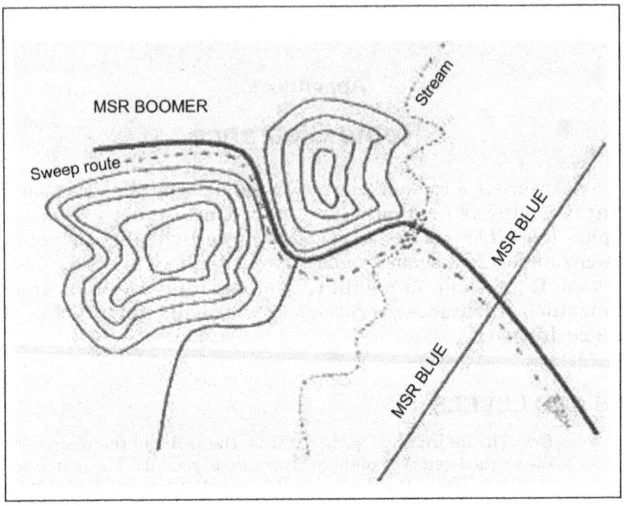

Figure E-1. Linear Clearance Method

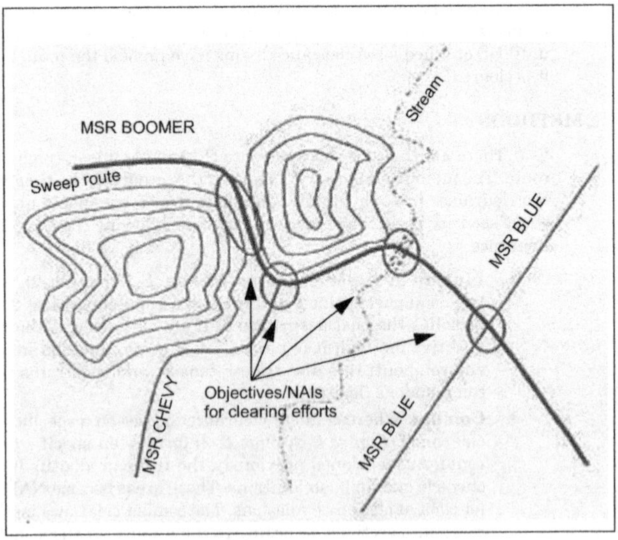

Figure E-2. Combat Clearance Method

that the route is secure and, if necessary, the sweep element can sweep the surrounding area if a minefield is found. Following the seizure of these objectives, the commander must assume a moderate risk that the S2 and the force engineer have identified all high-threat areas and that the route is clear of obstacles.

- **Combat route**. This method combines the clearance capabilities of the linear method with the security and surprise elements of the combat method. It is a two-phase, force-intensive operation and may require a battalion-size element, depending on the route's length. First, identified high-threat areas are targeted, secured, and cleared of obstacles and enemy forces before a sweep team moves. The sweep team then travels down the road and clears obstacles missed or not identified during the IPB/engineer battlefield assessment (EBA). The advantage of this method is that the TF commander immediately secures MSRs and then finds the enemy, confident that MSRs are relatively safe.

SWEEP LEVELS

E-4. The four levels of sweep operations are discussed below. All four levels can be modified to meet the unit's time and equipment limitations, but the commander assumes greater risk when the clearance level is modified. All four sweep levels can be used with any of the three clearance methods (linear, combat, and combat route).

- **Level one**. This is the fastest, riskiest type of route sweep. It is ideal for an armored/mechanized team. The sweep relies primarily on visual detection, using thermal or infrared sights or the naked eye for mine identification. It is followed immediately by a secondary, mechanical detection system such as MCRs. The sweep team is a squad-size or larger element that is task-organized with mine detectors, demolitions, and possibly an interim vehicle-mounted mine detector (IVMMD). The sweep team looks for minefields along the road width of a route. Security and support teams consist of a maneuver platoon to provide overwatching fire. The primary objective of a level one sweep is speed, with the sweep team moving at 5 to 8 miles per hour. The sweep team identifies immediate risks to traffic, neutralizes those risks, and continues on with the mission. If mine rollers are not available, a sandbagged 2 1/2-ton truck driven backwards can be used for a level one sweep, but only as a last resort. In this situation, the mine rollers are only a means of detection, not breaching. Mine rakes or plows are not always satisfactory substitutes because they most likely will destroy road surfaces.
- **Level two**. A level two sweep uses electronic measures as the primary detection method in high-threat areas. These areas include intersections, choke points, and areas within 10 meters of wood lines. A level two sweep employs more caution and forces the unit to update the IPB before beginning the mission. This level of operation employs a company team for security and C^2. Dismounted forces clear and secure the flanks and the far side of an identified minefield, while an engineer squad clears the road area.

- **Level three**. A level three sweep is more in-depth and more time-consuming. The sweep team must examine the route's entire width, including the shoulders and ditches. This technique ensures that follow-on forces are protected if they are forced to the side of the road. The security and support element also moves, mounted or dismounted, to provide rapid response and security. The sweep team provides a route recon report that updates current maps and further identifies high-threat areas. Staffs gain valuable information from this report for future operations.
- **Level four**. A level four sweep is very time-consuming. It relies on visual and electronic means for clearance. The platoon-size sweep team is dismounted to focus attention on the route's entire length. The company-size security element clears and secures the flanks and front of the sweep team (at least 100 meters in each direction in forested areas). This clears the area of off-route and command-detonated mines and allows the team to focus solely on the route. MCRs are used to proof the route after the sweep team passes through the area. Level four sweeps include route and area recons, and they are used when thoroughness and security, not speed, are critical to the mission.

EQUIPMENT

E-5. The IVMMD and the MinFlail are used to conduct route clearance.

INTERIM VEHICLE-MOUNTED MINE DETECTOR

E-6. The IVMMD (*Figure E-3*) is used in all levels of conflicts and stability and/or support actions. It is mounted on a blast- and fragmentation-survivable vehicle and is designed to detect and mark buried and surface-laid, metallic AT mines. The primary mission of the IVMMD is to detect mines during route-clearance operations. The system should not be operated in an environment where the enemy employs mines that are not pressure-fused.

System Components

E-7. A complete IVMMD consists of one mine-detection vehicle (MDV), one towing/mine-detection vehicle (T/MDV), three mine-detonation trailers, a spare-wheel module for the MDV, a spare-wheel module for the T/MDV, and a container of spare parts.

E-8. The MDV's only mission is to detect mines. It can negotiate vertical slopes up to a 20 percent grade. The MDV employs a 4-cylinder engine and a manual transmission to propel the 4.8-ton vehicle with a 3-meter-wide detection array. The detection array consists of two separate induction coils (one for the left side and one for the right side) that detect magnetic fields below the vehicle. The detection array is suspended between the two axles of the vehicle. When the detector encounters a metallic object, the operator is notified by an audible signal in the earphone and a visual signal on the dashboard that denotes which side array detected the object. The operator then stops the vehicle, backs it up, and reencounters the metallic object. (The MDV has two detection modes—the locate mode is used to identify the metal object, and the pinpoint mode is used to find the center of the object.) When

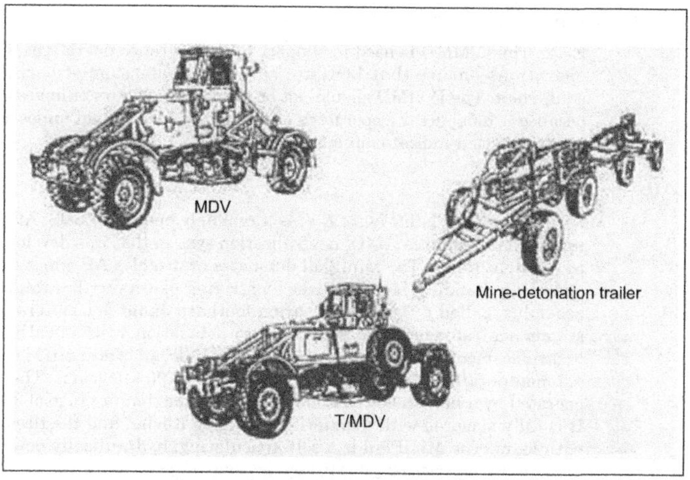

Figure E-3. IVMMD Components

the operator encounters the strongest signal, he activates the marking system (a nozzle mounted on the rear frame and centered on each detection array) that deploys a water-based ink onto the roadway.

E-9. The MDV is designed to produce very little ground pressure; however, it may detonate pressure-fused mines, depending on their fuse sensitivity. The MDV produces 27.9 pounds per square inch (psi) of ground pressure when the tires are inflated to 14.5 psi and 21.8 psi of ground pressure when the tires are inflated to 8.7 psi.

E-10. The T/MDV has a 6-cylinder engine and the same detection and marking system as the MDV. The T/MDV produces 49.8 psi of ground pressure when the tires are inflated to 29 psi and 23.4 psi of ground pressure when the tires are inflated to 8.7 psi. The T/MDV tows three mine-detonation trailers, and the recommended maximum operating speed while towing the trailers is 20 kph. The T/MDV (with trailers) can negotiate vertical slopes up to a 20 percent grade; however, going down such slopes is difficult. The T/MDV must be in first gear, and the trailer brakes must be deployed to decrease the speed of the vehicle when going down a slope.

E-11. The mine-detonation trailers are very heavy and specifically designed to apply heavy ground pressure that initiates pressure-activated mines. Each trailer has two axles of different lengths so that the three trailers provide a full 3-meter-wide proofing capability behind the T/MDV. If a mine detonates underneath the trailers, the wheel bolts are designed to sheer so that repair is limited to replacing a single wheel.

Employment Concept

E-12. The IVMMD is used to support route-clearance operations. Clearance operations ensure that LOC are safe for the passage of personnel and equipment. The IVMMD should not be used during hours of limited visibility, because it hampers the operator's ability to see surface-laid mines and visual signatures that indicate mining activities.

MINIFLAIL

E-13. The MiniFlail (*Figure E-4*) is a remotely operated, LOS, AP-mine and unexploded-ordnance (UXO) neutralization system that was developed for use by US light forces. The MiniFlail detonates or disables AP mines from a safe operating distance. It neutralizes by striking objects with a rotating chain assembly, called a flail, and clears a footpath about 1.1 meters wide. The system neutralizes AP mines and UXO by detonation, mechanical destruction, or displacement from the cleared lane. The MiniFlail is about 1.3 meters wide, 1.3 meters tall, and 3 meters long; it weighs 1,100 kilograms. The system is operated by a hand-held controller that has a maximum range of 300 meters. It is fully armored with a material similar to Kevlar, and the tires are filled with foam. The MiniFlail is a self-articulating, hydraulically powered shaft with 84 chains; each chain is 0.5 meter long.

Figure E-4. MiniFlail

PLANNING

E-14. The principles of breaching operations apply to developing and executing the route-clearance mission. The breaching tenets (intelligence, fundamentals, organization, mass, and synchronization) are the basis for planning.

INTELLIGENCE

E-15. Incorporating the IPB and METT-TC factors into route-clearance operations enables units to predict what the enemy will do and where it will do it. After the S2 and the engineer identify the most probable threat sites, the S2 designates them as NAIs. These NAIs are the focus of the recon effort.

Recon assets confirm the presence or absence of ambushes, UXO, and obstacles. The information gathered from the IPB and the recon effort determines the method and type of route clearance necessary. It also helps the commander determine any outside resources (explosive ordnance disposal [EOD] and special-operations forces [SOF]) that he may need.

FUNDAMENTALS

E-16. SOSRA may not be executed, but it is planned as it is in breaching operations. Units must be prepared to execute SOSRA fundamentals as necessary.

ORGANIZATION

E-17. Route-clearance operations are similar to breaching operations; therefore, commanders should plan and coordinate breaching fundamentals to suppress, obscure, secure, and reduce obstacles. Task-organizing for route clearance is also similar to that required for breaching operations. The assault force becomes the security element, the breach force becomes the sweep element, and the support force remains the same.

E-18. The significant difference between breaching and clearing operations is that breaching usually occurs during an attack (while under enemy fire) to project combat power to the far side of an obstacle. Route clearance focuses on opening LOC to ensure the safe passage of combat and support organizations within an AO. *Table E-1*, page E-8, shows a sample task organization for a route clearance.

MASS

E-19. Sufficient maneuver and engineer assets must be allocated to the sweep force. The length and width of the route and the type of clearance to be conducted determine the composition of the breach force. Clearing a Class A military road with the deliberate sweep technique requires at least two IVMMDs and/or two engineer squads due to the total lane width to be cleared and the requirement for rotating mine-detector operators. Depending on the type of sweep operation, the commander should plan for a 50 percent loss of sweep assets.

SYNCHRONIZATION

E-20. All aspects of synchronization should be implemented when planning route clearance. It is especially important that the clearance force conduct rehearsals at the combined-arms level. Rehearsals should include—

- Reaction to enemy contact, including ambushes.
- Communications exercise.
- Fire support (obscuration smoke, immediate suppression fires, CFZs, and no-fire area around the clearance site).
- CSS (casualty evacuation, marking materials, and demolitions resupply).

Table E-1. Sample Task Organization for a Route Clearance

Team	Support Force	Security Force	Sweep Force
Heavy	Mechanized infantry platoon with dismount capability Armor platoon FIST/COLT/Striker ADA section	Mechanized infantry platoon Engineer squad Mortar section Medical team (two ambulances) PSYOP team FIST MP element	Engineer platoon with organic vehicles Armor platoon with plows and rollers
Light/Heavy	Two infantry platoons (light) FIST/COLT/Striker ADA section	Bradley platoon with dismount capability Engineer squad 60-mm mortar section Medical team (two ambulances) PSYOP team FIST MP element	Engineer platoon with organic vehicles Armor platoon with plows and rollers
Light	Two infantry platoons (light) FO/COLT/Striker ADA section	AT/MP section with M60/MK19 mix 60-mm mortar section Medical team (two ambulances) PSYOP team FO MP element	Engineer squad (+) Infantry platoon (light) AT/MP section with M60/ MK19 mix

PLANNING CONSIDERATIONS

E-21. The purpose of breaching is to project combat power to the far side of an obstacle, and breaching usually occurs under enemy fire. Route-clearance operations focus on opening and maintaining LOC to ensure the safe passage of combat, CS, and CSS organizations. Like breaching, route-clearance operations require extensive Battlefield Operating System (BOS) coordination. The planning considerations shown in *Table E-2* should be used by brigade and battalion TFs when planning route-clearance operations. The obstacle template produced from this process determines which method and type of route clearance is required.

Table E-2. Planning Considerations for Route Clearance

INTELLIGENCE
Identify choke points, bridges, tunnels, critical road junctions, and built-up areas. These are the most likely locations for obstacle emplacement. However, depending on the enemy's overall mission, it may not always emplace obstacles at these locations. This is especially true if the enemy's goal is to psychologically disrupt convoys.
Maintain a situation map that reflects the most current intelligence information.
Maintain an incident map with a graphics overlay to facilitate a pattern analysis.
Maintain a threat order-of-battle database, such as how the enemy has disrupted LOC in the past.
Develop a detailed ISR plan that incorporates modern battle-space techniques and systems, such as ground sensors, forward-looking airborne radar, and satellite images. As a minimum— • Coordinate for UAV support, if available. • Develop infiltration routes to support recon and security at likely enemy ambush sites. • Develop an estimate of the impact to COBs (local nationals, NGOs, and PVOs). • Conduct a daily flight over the area to provide up-to-the-minute intelligence. • Coordinate with the USAF to periodically check the route (for example, use an AC-130 Specter gunship).
Establish liaison with the host nation, NGOs, and SOF.
Provide detailed OBSTINTEL, including— • Description of mines or explosive devices most likely encountered. • Composition and pattern of obstacles. • Enemy actions or techniques used during obstacle emplacement.
Provide intelligence updates on mine-hazard areas to company team leaders before departure. This should be in the form of a 1:50,000 enemy SITEMP overlay (confirmed and suspected/templated).
MANEUVER
Provide personnel for TCPs.
Secure flanks (at least 100 meters) and the far side of suspected and known obstacle locations.
Close the route to US-controlled traffic during route-clearance operations to minimize the target presented to enemy forces.
Identify and clear potential sniper positions before beginning obstacle reduction or clearance.
Provide security for the cleared route.
Plan the building of static security points along the cleared route to reduce the probability of reseeding.
FIRE SUPPORT
Plan smoke for templated obstacle locations.
Position mortars to ensure continuous coverage of the operation (one technique is to move under the control of the support force).
Prepare fires within the tactical rules of engagement.
Designate obstacle clearance sites as CFZs and no-fire areas.
Plan suppression fires on enemy elements capable of placing direct or indirect fires on the points of breach.

Table E-2. Planning Considerations for Route Clearance (continued)

MOBILITY/SURVIVABILITY
Report, clear, and mark obstacles and explosive devices to facilitate unimpeded movement.
Ensure that lane marking meets the standards directed by the supported commander and that materials and patterns are standard throughout the route.
Consider including road-repair equipment and material as part of the breach force (for example, a 5-ton dump truck filled with soil and an ACE to spread the soil).
Keep all radios, electronic equipment, and aviation assets at a safe distance during reduction operations.
Block uncleared roads and trails that branch from the route being cleared. This protects units from inadvertently traveling an uncleared route.
Debrief the chain of command and the TF S2 on the location, composition, and orientation of all obstacles cleared and encountered. This assists the S2 in IPB pattern analysis.
Provide detailed OBSTINTEL on minefields, including— • Description of mines or explosive devices most likely to be encountered. • Composition and pattern of obstacles. • Enemy actions or techniques used during obstacle emplacement.
ADA
Assist the S2 with the threat-air portion of the IPB.
Use the following passive air-defense measures: • Eliminate glare by using mud, tape, cardboard, or camouflage nets to cover headlights, mirrors, and portions of windshields. • Reduce dust clouds by reducing speed. • Plan routes that offer natural concealment. • Use air guards.
Increase the distance between vehicles.
Incorporate Stinger missile teams into the support force.
CSS
Ensure that clearance operations are supported by a logistical/CHS package.
Plan for air and ground evacuation of casualties. The preferred evacuation method is by air; the routine method is by ground.
Conduct an air-mission brief with air-ambulance assets, including pickup zones and markers.
Rehearse procedures for evacuation requests.
Ensure that the medical team consists of one or two ambulances.
Locate the medical team with the support force.
Identify the ambulance exchange point along the route to be cleared.
Ensure that appropriate personnel wear flak vests or IBASIC (*Figure E-5*, page E-12).
Ensure that all vehicles have tow cables in the front and the rear, and plan recovery assets for extraction purposes.
Ensure that all vehicles carrying troops have hardening (sandbags on floors and sides).
Provide MP and explosive-sniffing dogs to help in clearance operations and provide security for convoys during and after clearing operations.

Table E-2. Planning Considerations for Route Clearance (continued)

C^2
Designate a controlling, coordinating, and supporting HQ for route movement.
Designate, recognize, and include the following minefield indicators as part of company-team rehearsals: • Dead animals. • Avoidance by the local population. • Signs of digging. • Signs of concrete removal. • Disturbances in a road, such as holes or grooves. • Boxes or parcels placed along the road/shoulder. • Parked vehicles or bicycles without operators. • Wire on the road surface or extending onto the shoulders. • Evidence of vegetation disturbance along shoulders. • Evidence of mine-peculiar supplies such as wrenches, shipping plugs, wrapping paper or safety collars from fuses. • Posted signs that covertly alert the local population to the presence of mines. • Disturbances in tire tracks.
Designate a reserve force (at least platoon-size) that is mechanized or air-assault capable.
Ensure that proper rehearsals are planned and conducted. As a minimum, the unit should rehearse actions at the obstacle, actions on enemy contact, casualty evacuation, and control of COBs.
Ensure that the unit has a clear understanding of the mission, the intent, and the end state. For example, the clearing-unit commander should understand that his unit must clear the road width, including the shoulders, and secure the route.
Ensure that the unit— • Controls the movement of all personnel and equipment along the route. • Clears personnel within the minimum safe distance when clearing mines. • Prepares a risk assessment of the mission before issuing the OPORD. • Tracks the status of routes based on the amount of time since the route was cleared and the intelligence and enemy situations. • Tracks the progress of the clearance operation and integrates it into the maneuver and CSS plans. • Determines the route length, using clearly definable start and end points. • Coordinates with adjacent units, the host nation, NGOs, PVOs, and SOF. • Sets priorities for the route-clearance element.
SPECIAL OPERATIONS
Ensure that PSYOP/CA support the counterintelligence effort by conducting civilian interviews.
Direct civilians along the MSR to the displaced-personnel holding areas or along the routes that the unit has indicated for use.
Employ PSYOP/CA teams forward to disperse civilians and provide traffic management to isolate the route during clearance operations.

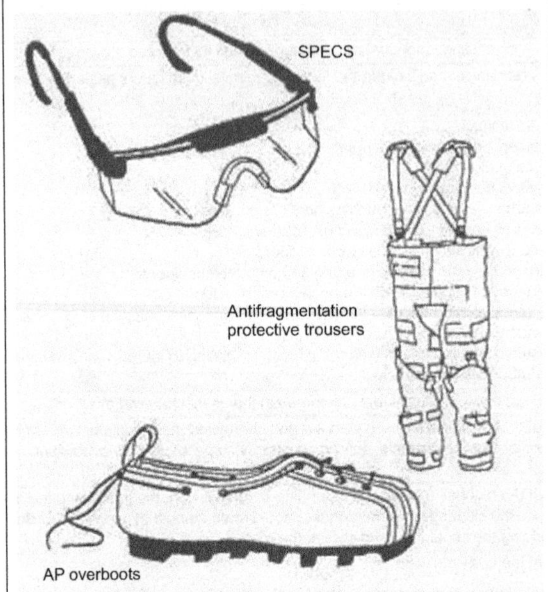

Figure E-5. IBASIC

TASK ORGANIZATION

E-22. A brigade or battalion TF normally conducts clearance-in-zone operations. To clear a route, the battalion TF focuses a company team as the main effort on the proposed MSR. *Table E-1*, page E-8, shows a sample task organization for a route clearance.

SUPPORT FORCE

E-23. The support force provides security for the breach force and neutralizes hostile forces. In rugged terrain or highly mined areas, moving the support force on the flanks may be too risky. In this situation, aviation assets may be used to provide flank security while ground forces provide rear security. The support force may also be tasked to search for suspected off-route mines.

SECURITY FORCE

E-24. The security force is prepared to maneuver and destroy enemy forces that the support force cannot neutralize. The security force may be comprised of maneuver and engineer forces augmented with a medical team, a psychological operations (PSYOP) team, and an EOD team. Additionally, the security force may be supported by a mortar section.

SWEEP FORCE

E-25. The sweep force is comprised of maneuver and engineer forces with detection equipment. The sweep force sweeps the route and reduces mine and explosive threats. As in breaching operations, the sweep force consists of a security element and a reduction element. The security element provides local security for the sweep force. The reduction element consists of sweep teams (personnel and equipment) to reduce obstacles and clear mines.

E-26. A sweep team is a trained detection team that searches for mines and explosive devices. The organization of the sweep team depends on the type of sweep mission, the length and width of the road to be cleared, the time and equipment available, and the surface composition (pavement, gravel, dirt) of the road to be swept. A dismounted platoon-size element can normally clear a 4.5-meter-wide path, and a squad-size element can normally clear a 1.5-meter-wide path. Mounted detection equipment, such as the IVMMD, can detect mines along a 3-meter-wide path. See *FM 20-32* for detailed information on the composition of a sweep team.

Glossary

AA avenue of approach

ACE armored combat earthmover

AD armored division

ADA air-defense artillery

ADAM area-denial artillery munition

AHD antihandling device

ammo ammunition

AO area of operations

AP antipersonnel

APC armored personnel carrier

APOBS Antipersonnel Obstacle Breaching System

ASLT assault

AT antitank

atk attack

ATTN attention

AVLB armored vehicle-launched bridge

AVLM armored vehicle-launched MICLIC

BFV Bradley fighting vehicle

BHL	battle handover line
BOS	battlefield operating system
C^2	command and control
CA	civil affairs
CAS	close air support
CCIR	commander's critical-information requirements
cdr	commander
CFZ	critical friendly zone
CHS	combat health support
CLAMS	Cleared-Lane Marking System
cm	centimeter(s)
cmd	command
COA	course of action
COB	civilians in the battle space
COLT	combat observation lasing team
CP	checkpoint
CS	combat support
CSS	combat service support
Czech	Czechoslovakia
DA	Department of the Army
det	detonated

dia diameter

DPICM dual-purpose improved conventional munition

DS direct support

DST decision-support template

ea each

EA engagement area

EBA engineer battlefield assessment

EOD explosive ordnance disposal

EOM end of mission

FD firing device

FECS Field-Expedient Countermine System

FFIR friendly forces information requirements

FIST fire-support team

FM field manual

FM frequency-modulated

FO foward observer

FRAGO fragmentary order

FSE fire-support element

FSO fire-support officer

FWF former warring faction

GPS Global Positioning System

HE high explosive

HEMMS Hand-Emplaced Minefield Marking System

HEMTT heavy expanded mobility tactical truck

how howitzer

HQ headquarters

IBASIC Improved Body Armor System, Individual Countermine

ICP intelligence-collection plan

ID infantry division

IDA improved dog-bone assembly

IED improvised explosive device

IFV infantry fighting vehicle

IP improved plow

IPB intelligence preparation of the battlefield

IR information requirements

ISR intelligence, surveillance, and reconnaissance

IVMMD interim vehicle-mounted mine detector

km kilometer(s)

kph kilometers per hour

lb pound(s)

LD line of departure

ldr leader

LO liaison officer

LOC lines of communication

LOS line of sight

m meter(s)

MCAP mine-clearing/armor-protection

MCB mine-clearing blade

MCR mine-clearing roller

MCRP Marine Corps reference publication

MDMP military decision-making process

MDV mine detection vehicle

METT-TC mission, enemy, terrain, troops, time available, and civilian considerations

MIBN motorized infantry battalion

MIC motorized infantry company

MICLIC mine-clearing line charge

MIP motorized infantry platoon

MLC military load class

mm millimeter(s)

MOUT military operations in urbanized terrain

MP military police

MSR main supply route

NA not applicable

NAI named area of interest

NATO North Atlantic Treaty Organization

NBC nuclear, biological, chemical

NGO nongovernment organization

NSN national stock number

OBJ objective

OBSTINTEL obstacle intelligence

OP observation post

OPFOR opposing force

OPORD operation order

oz ounce(s)

PCI precombat inspection

PIR priority intelligence requirements

PL phase line

psi pounds per square inch

PSN position

PSYOP psychological operations

PVO private volunteer organization

R&S reconnaissance and surveillance

RAAM remote antiarmor mine

RCU remote control unit

RL release line

RP reference point

S2 Intelligence Officer (US Army)

S3 Operations and Training Officer (US Army)

SBF support by fire

SCATMINE scatterable mine

SITEMP situation template

SOF special-operations forces

SOP standing operating procedures

SOSRA suppress, obscure, secure, reduce, and assault

SP start point

SPECS Special Protective Eyewear Cylindrical System

SPOTREP spot report

STANAG Standardization Agreement

T/MDV towing/mine-detection vehicle

TAI targeted area of interest

TC tank commander

TCP traffic-control post

TF task force

TM technical manual

TOC tactical operations center

TRADOC United States Army Training and Doctrine Command

TRP target reference point

TTP tactics, techniques, and procedures

UAV unmanned aerial vehicle

UK United Kingdom

UMCP unit maintenance collection point

US United States

USAF United States Air Force

UXO unexploded ordnance

WLGH weapon-launched grapnel hook

WO warning order

wt weight

Yug Yugoslavia

Bibliography

DA Form 2028. *Recommended Changes to Publications and Blank Forms.* 1 February 1974.

FM 3-50. *Smoke Operations.* 4 December 1990.

FM 5-71-2. *Armored Task-Force Engineer Combat Operations.* 28 June 1996.

FM 5-71-3. *Brigade Engineer Combat Operations (Armored).* 3 October 1995.

FM 5-71-100. *Division Engineer Combat Operations.* 22 April 1993.

FM 5-105. *Topographic Operations.* To be published in the next six months.

FM 5-170. *Engineer Reconnaissance.* 5 May 1998.

FM 6-71. *Tactics, Techniques, and Procedures for Fire Support for the Combined Arms Commander.* 29 September 1994.

FM 6-121. *Tactics, Techniques, and Procedures for Field Artillery Target Acquisition.* 25 September 1990.

FM 7-10. *The Infantry Rifle Company.* 14 December 1990.

FM 17-98. *Scout Platoon.* 9 September 1994.

FM 20-32. *Mine/Countermine Operations.* 29 May 1998.

FM 34-130. *Intelligence Preparation of the Battlefield.* 8 July 1994.

FM 71-1. *Tank and Mechanized Infantry Company Team.* 26 January 1998.

FM 71-2. *The Tank and Mechanized Infantry Battalion Task Force.* 27 September 1988.

FM 71-3. *The Armored and Mechanized Infantry Brigade.* 8 January 1996.

FM 90-7. *Combined Arms Obstacle Integration.* 29 September 1994.

FM 100-5. *Operations.* 14 June 1993.

FM 100-7. *Decisive Force: The Army in Theater Operations.* 31 May 1995.

FM 101-5. *Staff Organization and Operations.* 31 May 1997.

FM 101-5-1. *Operational Terms and Graphics (MCRP 5-2A).* 30 September 1997.

STANAG 2036. *Land Mine Laying, Marking, Recording and Reporting Procedures.* Edition 5. 28 January 1999.

TM 5-6665-298-10. *Operator's Manual for AN/PSS-12 Mine Detecting Set (TM 5-6665-298-10).* 28 February 1995.

TM 9-1375-215-14&P. *Operator's, Unit, Direct Support and General Support Maintenance Manual (Including Repair Parts and Special Tools List) for Demolition Kit, Mine Clearing Line Charge (MICLIC)*. 31 January 1992.

Index

FIELD MANUAL
No. 7-85

HEADQUARTERS
DEPARTMENT OF THE ARMY
Washington, DC, 9 June 1987

FM 7-85
RANGER UNIT OPERATIONS

IMPORTANT

U.S. Army Infantry School Statement on
U.S. NATIONAL POLICY CONCERNING ANTIPERSONNEL LAND MINES

TABLE OF CONTENTS

 DISTRIBUTION RESTRICTION: Approved for public release; distribution is unlimited.

PREFACE

This publication contains doctrinal guidelines for the employment and training of ranger units. The intended audience is intermediate and higher level commanders, staffs of those headquarters, and newly assigned officers and NCOs of the ranger regiment. It is to be used with all other published US doctrine concerning NBC operations and the climatic or geographic influences on combat. The fundamentals apply to the employment of ranger units on worldwide operations and across the spectrum of conflict from low to high intensity warfare. This includes all aspects of the integrated battlefield: nuclear, biological, and chemical. It is not intended to apply to the employment of long-range reconnaissance patrol (LRRP) companies, even though such units carry the designation "ranger." These LRRP companies, as well as the long-range surveillance company (LRSC) (corps level) and the long-range surveillance detachment (LRSD) (division level), are organized, trained, and equipped for long-range reconnaissance in enemy territory.

The scope of this publication addresses the mission, organization, equipment, capabilities, limitations, planning, training, operations, and logistical support of ranger units. It describes how the ranger regiment and its subordinate battalions train, receive taskings, plan missions, task-organize, deploy, and integrate supporting elements to perform the mission. Once in combat, the ranger regiment maneuvers to accomplish the mission in the manner of other light infantry units. Ranger units make full use of published light infantry doctrinal guidance. This manual will repeat procedures or techniques contained in other publications only if necessary for understanding or continuity.

This publication does not contain everything the ranger needs to know to successfully execute a unit mission. Rather, it supplements unit training and the ranger unit commander's guidance; it does not supersede it.

The proponent of this publication is HQ, TRADOC. Submit changes for improving this publication on DA Form 2028 (Recommended Changes to Publications and Blank Forms) and forward it to Commandant, US Army Infantry School, ATTN: ATSH-B-ID, Fort Benning, GA 31905.

Unless otherwise stated, whenever the masculine gender is used, both men and women are included.

CHAPTER 1

Introduction

The ranger regiment is a major component of the US Army's special operations forces. It is a unique light infantry unit tasked to conduct special military operations in support of national policies and objectives. These operations require highly trained, well-disciplined units capable of employment in any environment, either alone or in concert with other military forces. Within this publication, the term *ranger force* describes any size force consisting mainly of members of the ranger regiment and led by a member of the ranger regiment's chain of command. A ranger force may be a TOE unit or it may be a specially organized task force for a specific mission.

1-1. Mission.

a. The mission of the ranger regiment is to plan and conduct special military operations. These operations are conducted by specially trained, equipped, and organized forces against strategic or tactical targets in pursuit of national military, political, economic, or psychological objectives. They may support conventional military operations or they may be performed independently when conventional forces cannot be used. (See Figure 1-1.)

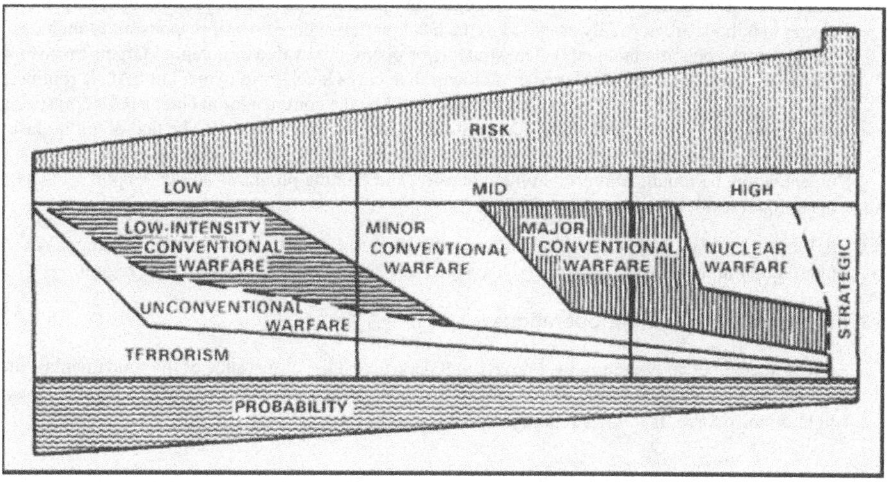

Figure 1-1. Spectrum of conflict.

b. Special military operations conducted by the ranger regiment include strike operations, usually deep penetration, and special light infantry operations. Strike operations include raids, interdiction, and recovery operations. Special light infantry operations include many of the light infantry missions assigned to airborne, air assault, or light infantry battalions and brigades. These operations are conducted in support of the AirLand Battle at all levels of intensity.

1-2. Purpose and function.

The ranger regiment provides the national command authority (NCA) the ability to move a credible military force quickly to any region in the world. The regiment uses the entire spectrum of intelligence support, from national systems to organic assets. Ranger units maintain a readiness posture that supports their immediate commitment to battle once deployed. They are often tailored for specific missions and may require augmentation from external sources. Tactical mobility may be augmented by USAF or Army special operations aviation (SOA) aircraft.

a. The ranger regiment conducts light infantry operations on the integrated battlefield as well as in low intensity conflicts. Special light infantry operations are conducted to accomplish certain contingency missions during peacetime or as part of the echelons above corps (EAC) battle.

b. The regiment conducts both strategic and operational missions. Its efforts are combined into an overall plan to destroy, delay, and disorganize the enemy, or to cause him to divert his attention and combat forces to rear area security. The limited number of ranger units and the diverse targets dictate a careful assignment of missions.

c. The mission, enemy, terrain, troops and time available (METT-T) factors govern the command and control arrangement under which the regiment operates. The ranger regiment or separate ranger battalions are normally assigned to the headquarters whose area of responsibility includes the regiment's operational area(s). The strategic or operational value of potential targets means that ranger units are normally employed at no lower than corps level. Employment at EAC is routine. The EAC headquarters employing ranger units could be the commander in chief's (CINC's) special operations command or field army. The controlling headquarters could also be that of a joint task force (JTF) or the Army component command (ARFOR) of a JTF. The ranger regiment's organization, communications equipment density, and training programs do not support operational level reconnaissance missions.

d. Ranger battalions are not oriented to a specific theater. Current force structure, contingency plans, and training needs (see Appendix G) preclude committing battalions to one region.

1-3. Fundamentals of ranger operations.

a. The success of an operation by a ranger unit depends on the observance of the fundamentals of the US Army's AirLand Battle doctrine (see FM 100-5). The ranger regiment fights the enemy as a light infantry force. It follows infantry doctrine and observes all the basic rules of infantry operations.

b. Along with initiative, depth, agility, and synchronization, ranger operations require--

(1) Detailed planning and coordination that allow the ranger unit to discern and exploit the enemy's weaknesses while avoiding his strength. Both permit the evaluation and use of information gathered from all sources. Detailed coordination integrates all supporting units and services. During planning, the ranger unit conducts pre-mission training, briefings, and rehearsals for all personnel. The use of special mission equipment is practiced and perfected.

(2) Decentralized execution IAW mission orders and the commander's intent. Special operations forces must use individual and unit initiative. Mission-type orders give the ranger force commander the flexibility to take advantage of opportunities on the battlefield.

(3) Surprise, achieved through the ranger unit's ability to move by uncommon means, along unexpected routes, and over rough terrain. Ranger units normally conduct operations during poor weather and reduced visibility, aided by night vision devices. This adds to the attainment of surprise.

(4) Survivability, achieved by using the classic infantry combat techniques of stealth and concealment. The ranger unit engages the enemy at the time and place of its own choosing. It takes full advantage of terrain, and destroys or suppresses enemy weapons. The ranger unit seeks to destroy or neutralize the enemy's command and control systems, his surveillance assets, and his mobility assets. Survivability is enhanced by rapid mission accomplishment and a prompt departure from the objective area.

(5) Mobility, speed, and violence of execution that allow the ranger unit it to close quickly on the objective area and complete the mission before the enemy can react. The speed at which events take place confuses and deceives the enemy as to the intent of the ranger unit. This forces the enemy to react rather than to take the initiative. Tactical mobility lets the ranger unit break contact and withdraw from the objective area.

(6) Shock effect, which is a psychological advantage achieved by the combining of speed and violence with the precision of the ranger attack. The ranger unit strives to apply its full combat power at the decisive time and place, and at the point of the greatest enemy weakness. The ranger unit strives to achieve maximum physical and psychological effect on the enemy by exhibiting aggressiveness and reasoned audacity.

(7) Multiple methods of insertion and attack, trying not to repeat operations thus decreasing the chance the enemy will detect a pattern. This is achieved through imaginative training and planning.

(8) Deception, practiced by the ranger unit during all phases of its operations, from deployment through the insertion phase, to the actions in the objective area and extraction. The enemy is kept unaware of the ranger unit's intentions, is fed conflicting and wrong information, and is kept from knowing the size or mission of the ranger force. The ranger force commander makes full use of ruses, feints, false insertions, electronic countermeasures, and dummy transmissions. Concurrent military operations by US or allied forces may be used as cover for ranger operations. Ranger units are careful to hide every possible aspect of an operation and to disguise those that cannot be hidden. While in the operational area, or when departing any location, ranger units ensure that no material is left that could provide the enemy with information concerning the ranger force.

(9) Audacity, achieved by a willingness to accept risk. The ranger force commander considers what the enemy expects the unit to do and the actual ranger abilities, He then chooses a course of action that may confuse the enemy, while remaining within the capability of the ranger unit. This fundamental is combined with deception and surprise to disrupt the enemy command and control.

c. Detailed planning and coordination finds the enemy. Surprise, mobility and speed, variety, deception, and audacity combine to shock and disorient the enemy, fixing him in place. The violence and precision of the ranger attack finishes the enemy while ensuring the ranger force survives.

1-4. Capabilities and limitations.

a. The ranger regiment has the following capabilities:

(1) Deploying quickly to conduct operations on all types of terrain and in all kinds of weather.

(2) Establishing a credible American presence in any part of the world to show US interest or resolve.

(3) Infiltrating and exfiltrating an area of operations, and assaulting an objective by land, sea, or air.

(4) Conducting strike operations to include raids, personnel and equipment recovery operations, and interdiction of key areas.

(5) Conducting special light infantry operations to include seizing and securing airfields, communications centers, command and control facilities, and key bridges; and other special light infantry operations.

(6) Performing short-duration reconnaissance of assigned ranger objectives for the ranger force commander.

(7) Operating for up to three days without resupply, and for longer periods when provided with accompanying or airdropped supplies.

(8) Providing liaison, communication, and coordination personnel and equipment to integrate the deployed ranger force into the logistical, intelligence, and operational system of the theater or joint task force (JTF) commander.

(9) Assuming operational control, for a limited time, of other US military forces such as engineers or infantry, airborne, or air assault battalions.

(10) Providing the focal point for all-source intelligence support to attached and assigned units of the regimental task force.

(11) Conducting limited combat operations under conditions of chemical, nuclear, or biological contamination.

b. The ranger regiment has the following limitations:

(1) Limited capability against armored or motorized units in open terrain.

(2) No organic transportation.

(3) Limited sustained combat capability due to the shortage of organic combat support and combat service support elements.

(4) Limited organic air defense weapons.

(5) Limited organic indirect fire support.

(6) No casualty evacuation capability.

(7) Reconstitution and retraining needed to replace combat losses.

1-5. Employment considerations.

a. Ranger units are characterized by the quality, motivation, training, and individual skill of their members. This produces units with superb collective abilities, able to adapt well to changing, complex situations.

b. Ranger units can conduct either deliberate or quick-response operations.

(1) Deliberate operations rely on careful planning, reconnaissance and surveillance of the target area, deception, secrecy, thorough preparation and rehearsals, and violent execution. A deliberate operation aims to complete the mission even though the enemy may have heavy forces on or near the objective area. Deliberate operations allow for detailed planning, evaluation, rehearsal, and coordination before insertion. A deliberate operation is likely to succeed against targets that the enemy has protected in depth, that have strong natural defenses, or that need a detailed and long insertion process.

(2) Quick-response operations rely on the high level of training and readiness of the ranger regiment to execute a mission before the enemy can react. These operations are conducted when there is little time for long, detailed planning. They rely on set procedures set forth by the ranger regiment and its supporting elements. A quick-response operation aims to complete the mission before an enemy can react. This type of operation may be chosen due to the time-sensitive nature of the target, political or military goals, the time frame of other operations, or the increased chance of enemy detection.

(3) Whether an operation is to be deliberate or quick response is often a difficult and time-sensitive decision by a high-level command authority. Decision-makers must consider the enemy's strength in the area, his intentions, his ability either to reinforce or to alter the target area, and the consequences of success or failure of diplomatic or military initiatives in related areas. The ranger unit commander tries to use the existing time, manpower, and resources to complete a detailed and coordinated plan. He refines that plan up to the insertion into the objective area.

c. Ranger units train to operate in any environment or weather condition. They regularly perform operations during periods of limited visibility. Ranger units maintain a high state of physical fitness and often train in close quarters combatives. In addition to completing advanced marksmanship training with standard US weapons, each member of a ranger unit trains with many foreign weapons. Ranger units are trained to operate on urbanized terrain, becoming specialists in entry and clearing techniques and quick-fire methods, especially during periods of limited visibility.

d. Because ranger units have limited vehicles, logistics operations capability, indirect fire support, and heavy weapons systems, they are not designed for continuous operations. During all phases of operations and training, ranger units need responsive external support.

e. Ranger units are normally employed against targets and under conditions that need their unique skills. Although targeting priority is set by the overall commander, ranger units are not normally assigned missions that can be done by conventional aerial bombardment or by other units.

f. Ranger units are oriented toward offensive operations. They are not normally employed as a rear area protection force. Although the ranger regiment has a small reconnaissance unit, ranger units do not normally conduct long-range reconnaissance missions. The structure, communications, and training of the ranger unit do not prepare it for LRRP missions. Ranger units engaged in strike and special light infantry operations have a secondary mission to collect and report combat information.

g. Ranger units can be deployed worldwide when US military presence or participation with a host national military activity would serve US interests. This deployment shows a readiness to commit forces into a threatened area or proves US national resolve. After the deployment, other activities include staging operations, rehearsing combat operations, securing base areas for use and deployment of other forces, and so on that provide a clear signal of US intent. Ranger units are not trained or organized to provide mobile training teams (MTT) to train indigenous forces. The US Special Forces or other special operation forces are trained to conduct such security assistance operations.

h. Ranger units can serve as an example to a host country and provide limited military advice and training. They will normally be augmented with linguists and technicians to increase their abilities. The regiment would normally still function as a unit. Most of its assistance would be through short-term, high-impact, unit-oriented operations. They would not be long-term individual efforts associated with advisory-type activities.

i. Ranger units may be deployed to engage in combined training exercises with allies. This enhances US national image by demonstrating the outstanding abilities of the American ranger. These activities may include ranger, light infantry, airborne, air assault, or amphibious operations.

j. Ranger units use standard US nuclear, biological, chemical (NBC) warning; detection; protection; and decontamination equipment and doctrine when operating on the integrated battlefield.

1-6. Command and control.

a. **External.**

(1) The ranger regiment is a key component of the US Army's special operations forces. The other elements of special operations forces are special forces (SF), psychological operations (PSYOP), civil affairs (CA), and special operations aviation.

(2) The 1st Special Operations Command (1st SOCOM) is responsible for command of all active component special operations forces elements, to include the ranger regiment, in peacetime. The 1st SOCOM has command and control of all assigned special operations forces, less operational command (OPCOM) of special operations forces units forward deployed. (See Figure 1-2.) The ranger regiment, while based in the continental United States (CONUS), is under the control of 1st SOCOM. During peacetime, the OPCOM of deployed ranger units is through channels chosen by the NCA through the Joint Chiefs of Staff (JCS) to the unified command.

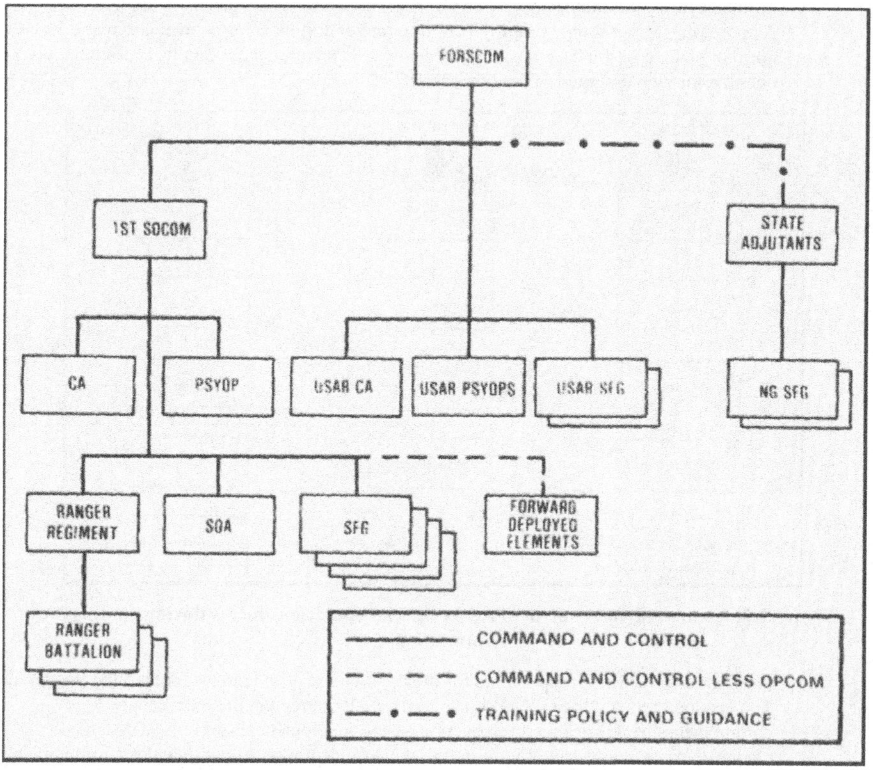

Figure 1-2. Command and control of Army Special operations forces during peacetime.

(3) In support of national policy, the NCA may direct contingency operations involving US

forces. These are politically sensitive military operations normally characterized by the short-term rapid projection or employment of forces in conditions short of conventional war. One aim of the NCA is to conclude them without going to war. The flexibility and effectiveness of the ranger regiment make it possible to use military force quickly to complete the mission and limit the spread of conflict. Command and control of special operations forces in a contingency operation is usually by a special operations task force (SOTF). (See Figure 1-3.) The SOTF and designated special operations forces are under the OPCOM of the highest level of command responsible for the operation. This command may be the unified command responsible for the contingency area (Option A), or a JTF designated by the NCA through the JCS (Option B). If no SOTF is formed, the ranger regiment could be OPCOM to the Army component commander (Option C). The ranger regiment or its battalions are not normally placed OPCOM to a division. A division's area of influence rarely contains appropriate targets. The division also lacks the assets to adequately support ranger operations. Employment of a ranger unit by a corps-level command is normally on a case-by-case basis. The theater or JTF commander retains centralized control over the mission assigned to the ranger unit.

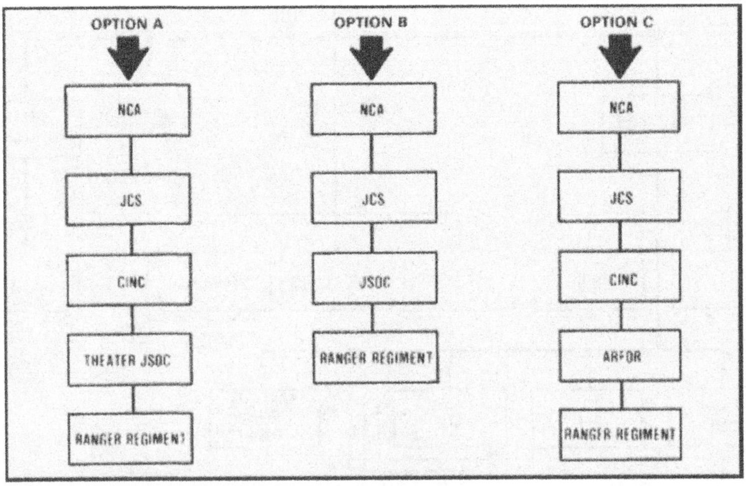

Figure 1-3. Command and control of Army Special operations forces during contingency operations.

(4) The 1st SOCOM is responsible for the deployment of the ranger force to the theater area. The headquarters having OPCOM provides the all-source intelligence, secure communications, insertion and extraction assets, and logistical support needed. The responsible commander in chief, through the Army component commander, provides support during the contingency operation.

(5) During wartime, ranger units are deployed as strategic assets to the theater of operations.

Strategic mission support is provided to the unified commander (CINC) from the NCA through the Joint Chiefs of Staff. It may involve the execution of sensitive operations. The employment of ranger units is controlled at the highest level of command directly responsible for the mission. Ranger units are assigned to a REMAB based on mission, force protection, and OPSEC considerations. The employing command and other elements provide timely and accurate intelligence, communications, logistical, and administrative support. Command and control of these units conforms to the provisions of JCS Pub 2. (See Figure 1-4.)

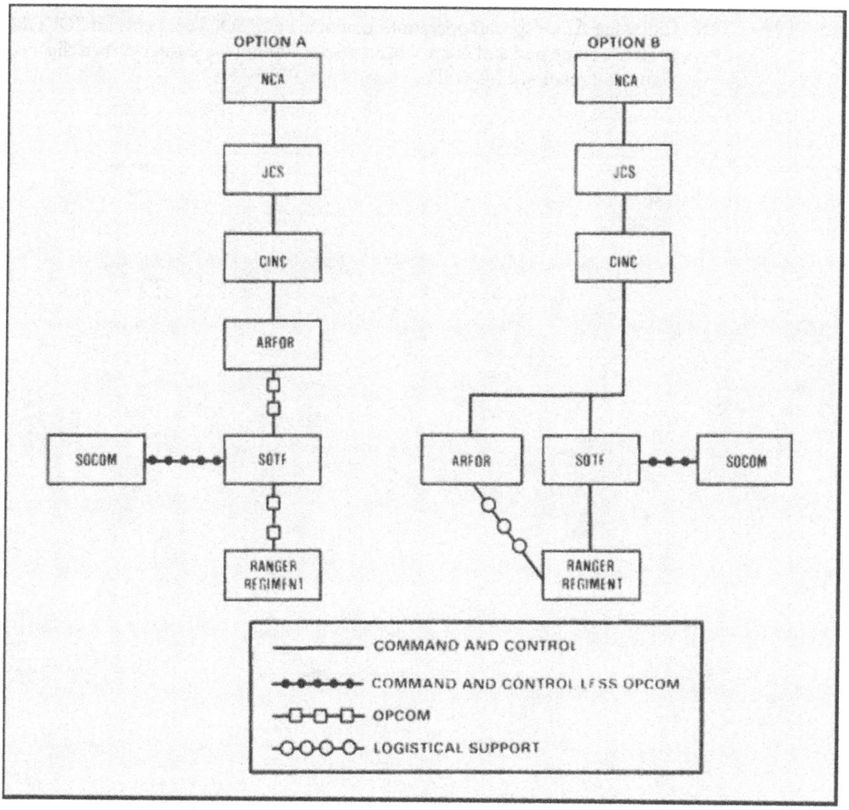

Figure 1-4. Command and control of Army Special operations forces during wartime.

(a) A theater joint special operations command (JSOC) is subordinate to the unified command to supervise the command and control, employment, and support of all US and allied special operations forces. The JSOC is a key element in the command and control of ranger units performing strike missions or special light infantry operations (Option A).

(b) The JSOC may be placed on a lateral line with other in-theater service components (Option B). This can occur when ranger units are -employed far from conventional forces and against strategic targets. It can also occur when the JCS or a CINC forms a JTF to conduct a limited-duration operation in an area where ranger units and conventional forces may operate near each other, or where operations must be coordinated on a regional level.

(c) Under the Army special operations command (ARSOC) concept, 1st SOCOM deploys as a command and control headquarters within the theater. When this occurs, the command relationships will be as shown in Figure 1-5.

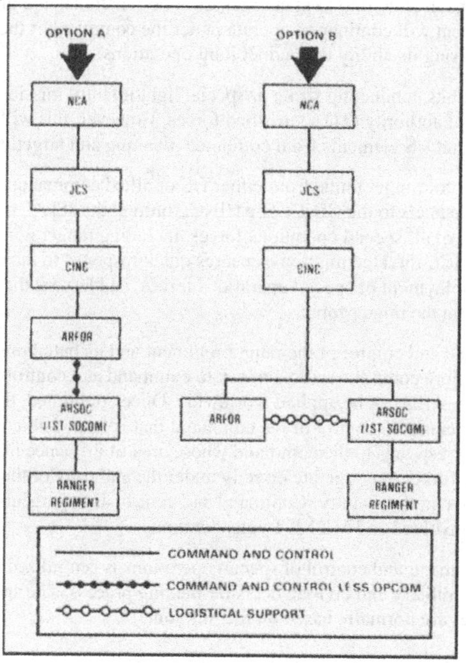

Figure 1-5. Command and control of Army Special operations forces under the ARSOC concept.

(d) Recently passed legislation will modify existing high level SOF command and control arrangements. All US Army, USN, and USAF special operations forces will be consolidated into joint commands at both national and theater level. The ranger regiment will continue to operate under the command of the 1st SOCOM, while increasing its ability to conduct joint operations.

(6) Ranger units conducting strike or special light infantry missions are not normally offered for transfer of authority (TOA) to allied forces. However, this will not preclude allied commands and US elements from combined planning and targeting for ranger missions.

(7) Requests for ranger units, from either US or allied commands, go through normal command channels to the CINC. The CINC, through the JSOC, is responsible for the employment of all special operations forces, including ranger units. This provides flexible response to JCS-directed missions, ensures quick response to the CINC's needs, enhances the joint employment of special operations forces, and lessens the chance of overcommitting and degrading the ranger force.

(8) Command and control of the ranger regiment and its battalions is IAW AirLand Battle doctrine. Before commitment to a mission, command and control is kept at a level where the unit's unique skills can be applied worldwide. Once committed, the ranger unit is placed under the operational control of the command that is responsible for the mission. This normally corresponds to the command whose area of influence includes the target area. Ranger task forces may operate directly under the authority of the senior American representative in the country. Command and control arrangements may vary according to the theater involved and METT-T considerations.

(9) The command and control of special operations is centralized and begins at the highest level of the military and civilian decision-making process. The specific command relationships are normally based on the mission.

b. **Internal.**

(1) The ranger regimental headquarters is organized to operate like that of a brigade headquarters. In addition to commanding and controlling all three ranger battalions, the regimental headquarters can assume operational control (OPCON) of conventional combat and combat support units, and other special operations forces for limited periods. It provides the regimental commander with a well-balanced responsive staff. It is structured as an operational headquarters that can deploy to the mission area and act as the ground tactical headquarters.

(a) The regimental headquarters prepares for combat by forming two small command and control groups. Each of these can deploy to an objective area and control combat operations. Other elements of the staff operate the tactical operations center (TOC) and the logistical operations center (LOC). These may be collocated depending on the situation. The TOC and LOC are normally located well behind the FEBA, at the REMAB or at the intermediate staging base (ISB). The ranger regiment also has two small liaison teams consisting of LNOS, staff representatives, and communications elements. (See Appendix B.)

(b) When a single ranger battalion is committed to an operation, the regimental commander normally exercises control from the TOC. A liaison and communications team is detached from the regimental headquarters to augment the ranger battalion. This team is attached to the deployed battalion until the end of the mission.

(c) If two or more ranger battalions are used in a single operation, the regimental commander normally deploys both command and control groups. He commands from the objective area, The use of two small, mobile control elements makes up for the possible loss of the regimental commander and the primary control element.

(d) Whenever the regimental commander is controlling an operation from the REMAB or the objective area, he normally provides a liaison team with a communications element to the higher headquarters exercising operational control. This cell can provide ranger representation to the controlling staff, operations interface, logistics interface, fire support planning and coordination, intelligence analysis and dissemination, and communications between the higher headquarters and the ranger battalion.

(2) The ranger regimental headquarters can serve as a ranger or special operations force task force headquarters. With augmentation, it can function as the Army special operations force command element. The regimental headquarters can also function as the Army component command of a JTF.

1-7. Communications.

a. Ranger missions need unique and dedicated communication support assets. Along with the ranger regiment's organic communications means, communications support assets are organized within the 1st SOCOM that-can deploy in support of the ranger regiment. This ensures that there are reliable communications to all elements of the SOTF. A special operations communications battalion (SOCB), deployed by the 1st SOCOM, provides more communications systems to support the special operations plan. The SOCB does not replace or duplicate existing in-theater communications. It augments the CINC's communications with the US Army's portion of the critical special operations force command, control, and communications link unique to the SOTF.

b. The ranger regiment's communication platoon can provide secure, long-range communication links from the CINC to a deployed ranger force. It can also enter the SOCB and the joint communications support element (JCSE) nets to provide a direct link with the NCA. Its equipment consists of four light vehicles equipped with various multiple communications systems. The two regimental liaison teams use one of these vehicles each. The TOC and LOC each use one.

c. Secure satellite communications (SATCOM) and radios with AM or FM capability are the main means of communication within the ranger regiment. These are also used for communications within the ranger battalions. The regimental commander has two light vehicles equipped with multiple communications systems. One of these is used to support each of the two mobile command and control elements of the regimental headquarters. Some of the radio systems can be dismounted and man-packed.

d. The ranger regimental headquarters has both man-packed and vehicular tactical satellite communication systems. Each of the ranger battalions has one base station and three man-packed systems.

(1) The man-packed radio is battery-operated. It provides two-way communications in both line-of-sight (mobile) and satellite (at-halt) modes. The digital message device group (DMDG) is the input and output device for data transmission. Secure voice communications is achieved using various encryption devices.

(2) The vehicular net control station uses the basic receiver/ transmitter unit installed on a vehicle. Each can serve up to 15 terminals in a communications net.

e. The reconnaissance platoon is equipped with multiple, man-packed communications systems. These let its teams operate in most of the communications nets within the regiment. Each five-man reconnaissance team has this communications ability, as does the platoon headquarters.

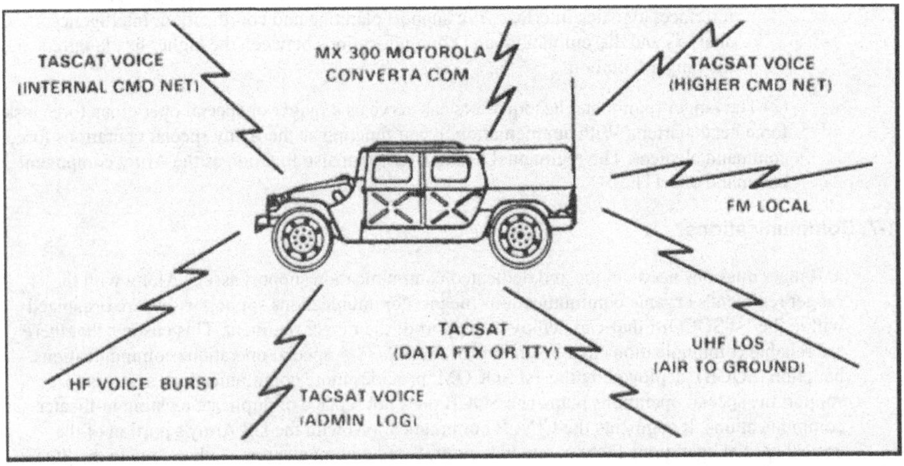

Figure 1-6. Communications capabilities within the regimental headquarters.

CHAPTER 2

Organization and Equipment

The ranger regiment is a specially organized and trained light infantry force consisting of about 2,000 personnel. It includes a regimental headquarters and headquarters company and three ranger battalions. (See Appendix A.) The regiment can operate as a single unit, separate battalions, or separate companies on specific missions. Independent missions are not normally assigned to ranger platoons.

2-1. Headquarters and headquarters company.

The HHC of the ranger regiment consists of 130 personnel, the majority being senior noncommissioned officers, commissioned officers, or soldiers with special skills to support the operations of the ranger battalions. (See Appendix A.) The HHC consists of two elements: the regimental headquarters and the headquarters company (minus).

2-2. Regimental headquarters.

a. The regimental headquarters consists of the regimental commander and his staff. The primary staff is composed of five sections: personnel (S1), intelligence (S2), operations and training (S3), supply (S4), and civil affairs (S5). These staff sections perform the standard functions and are supervised by the executive officer (XO). The regimental commander has a deputy commander and a special staff consisting of a communications officer, fire support officer, surgeon, and staff judge advocate. The S2 officer has a large intelligence analysis and counterintelligence section. The S3 section has extra liaison officers and assistant operations officers to coordinate the widespread and complex ranger missions. A United States Air Force (USAF) staff weather officer and a tactical air control party (TACP) are permanently attached.

b. The regimental headquarters:

- Analyzes assigned missions.
- Plans, coordinates, and controls training and operations.
- Coordinates the infiltration, resupply, exfiltration, and debriefing of the reconnaissance teams.
- Provides liaison support.
- Provides administrative, medical, NBC, and logistical planning and coordination support to the committed ranger battalions.
- Coordinates special support requirements (rigger, transportation, communications security [COMSEC]) for the ranger battalions.

2-3. Regimental headquarters company.

The regimental headquarters company (minus) consists of the company commander and his administrative personnel, plus the limited combat support (CS) and combat service support (CSS) elements assigned to the regiment. These are the fire support element, the communications platoon, the reconnaissance platoon, and the medical treatment team.

a. The company headquarters:

(1) Provides command and control of assigned elements.

(2) Provides administrative, personnel, logistical, and maintenance support to the assigned elements.

b. The fire support element:

(1) Plans fires to support the commander's ground tactical plan.

(2) Coordinates the integration of all available fire support means (mortars, artillery, armed helicopters, close air support, naval gunfire) to support mission execution.

c. The communications platoon (see Appendix A):

(1) Provides secure SATCOM teletype (TTY) and facsimile transmissions (FAX) abilities at the remote marshalling base or intermediate staging base (REMAB or ISB).

(2) Provides secure SATCOM, TTY, and FAX to the regimental headquarters and ranger battalions in the objective area.

(3) Provides cryptographic support to the regimental headquarters.

(4) Provides limited repair of radio and cryptographic equipment.

d. The reconnaissance platoon (see Appendix A):

(1) Trains, equips, and controls the employment of three reconnaissance teams.

(2) Infiltrates the objective area by parachute (high-altitude, low-opening [HALO]; high-altitude, high-opening [HAHO]; static line), helicopter, fixed-wing aircraft, self-contained underwater breathing apparatus (SCUBA), small boat, or other means.

(3) Remains undetected in the objective area for up to five days.

(4) Performs reconnaissance.

(5) Performs demolition target analysis.

(6) Uses small boats and inflatable rafts.

(7) Emplaces and monitors unattended ground sensors.

(8) Uses night image enhancement devices.

(9) Uses photographic equipment.

(10) Emplaces electronic target acquisition and designation devices.

(11) Collects specific combat information to satisfy priority intelligence requirements (PIR).

(12) Reports information using secure, long-range communication equipment.

Figure 2-1. Reconnaissance platoon communications.

(13) Performs drop zone (DZ) and landing zone (LZ) selection, marking, and reception duties.

(14) Escapes and evades enemy areas.

(15) Performs limited attacks or ambushes when tasked by the regimental commander.

(16) Reports weather in the objective area.

e. The medical treatment team:

(1) Provides medical support to the regimental headquarters.

(2) Coordinates required medical support and evacuation.

2-4. Ranger battalion.

a. The ranger battalion is organized and equipped to conduct both special and conventional combat operations. It contains organic combat, CS, and limited CSS elements. It consists of a battalion headquarters, a headquarters company, and three ranger rifle companies. (See Appendix A.)

b. The ranger battalion is the combat element of the ranger regiment. It can conduct independent combat operations or subdivide into its separate ranger rifle companies and their subordinate elements who can conduct semi-independent operations.

c. The ranger battalion headquarters and headquarters company (see Appendix A) includes:

(1) **Battalion headquarters.** The battalion headquarters consists of the commander; the executive officer; the command sergeant major; the S1, S2, S3, S4, and S5 officers; the chaplain; the communications officer; the fire support officer; and two radio operators.

(2) **Battalion headquarters section.** The battalion headquarters section consists of the S3 air and assistant S8, an assistant S2, an intelligence sergeant and a senior intelligence analyst, two operations sergeants and an operations assistant, a personnel staff noncommissioned officer (PSNCO), the chaplain's assistant, two clerk typists, two radio operators, a chemical officer and NCO, a personnel administration center (PAC) NCO, three PAC clerks, and a legal specialist. There is a USAF tactical air control party permanently attached.

(3) **Communications section.** The communications section consists of one telecommunications technician, one tactical communications chief, two combat signal radio team chiefs, one tactical communications systems supervisor, three radio operators, one switchboard operator, one single channel radio operator, and two equipment mechanics. The section provides communications support and associated maintenance.

(4) **Battalion medical section.** The battalion medical section consists of one general medical officer, one physician's assistant, two medical specialists, and twelve aidmen. The section provides unit-level medical support and trains selected members of the battalion in advanced medical procedures.

(5) **Battalion support platoon.** The battalion support platoon consists of a battalion support platoon leader, an ammunition NCO, and a food service section. The platoon provides limited CSS for assigned and attached units. This platoon does not have organic transportation. The battalion food service section consists of a food service sergeant (E7), a first cook (E6), and eight cooks.

(6) **Fire support team headquarters.** The fire support team (FIST) headquarters is supervised by the battalion fire support officer. It consists of the fire support sergeant and two fire support specialists. The headquarters controls three 9-man fire support teams. These teams are normally with the ranger rifle companies in a habitual support relationship. When operations are conducted near water, the fire support team headquarters may be augmented by spotter teams from the naval shore fire control party.

(7) **Company headquarters.** The HHC company headquarters consists of a commanding officer, an executive officer, a first sergeant, a supply sergeant, and an armorer.

d. The ranger rifle company (see Appendix A) includes a company headquarters, three rifle platoons, and a weapons platoon. Each company can conduct semi-independent operations for a short time when augmented by assets from the ranger battalion headquarters company.

(1) **Ranger rifle company headquarters.** The ranger rifle company headquarters consists of a commanding officer, an executive officer, a first sergeant, a supply sergeant, a tactical communications chief, two radiotelephone operators, an armorer, and an NBC NCO. It also has a two-man sniper team equipped with a sniper weapon. It normally has one of the 9-man FISTs attached from the ranger battalion headquarters company. Elements of this FIST may be further attached to the ranger rifle platoons.

(2) **Ranger rifle platoon.** The ranger rifle platoon consists of a headquarters section, three rifle squads, and a machine gun squad. The headquarters section personnel include a platoon leader, a platoon sergeant, and a radiotelephone operator. Each rifle squad consists of a squad leader and two balanced fire teams of four men each. The machine gun squad consists of a squad leader and three M60 machine gun crews of three men each. The machine gun squad is normally employed as a single element to provide quick, accurate, long-range supporting fire. One or more crews may be attached to other squads within the platoon.

(3) **Ranger weapons platoon.** The ranger weapons platoon consists of a headquarters section, a mortar section, and an antiarmor section. The headquarters section consists of a platoon leader, a platoon sergeant, and a radiotelephone operator. The mortar section has two 3-man 60-mm mortar squads. The mortar squads are normally employed as a section. They may be attached by section or squad to other platoons. The antitank section normally consists of a section sergeant and three 3-man 90-mm recoilless rifle or Dragon missile teams. They are normally employed as a section. The antiarmor section may also be attached by section or team to other platoons.

2-5. Special teams and elements.

The ranger regiment forms special teams to do critical tasks. These teams are made up of members of the unit and are used when needed. Examples of these special teams and elements are--

a. SCUBA teams whose members are graduates of the Special Forces Underwater Operations School. They conduct special insertion and reconnaissance duties.

b. HALO teams whose members are graduates of the Special Forces Freefall School and who conduct high-altitude, low-opening parachute operations.

c. HAHO teams whose members are also graduates of the Special Forces Freefall School and who conduct high-altitude, high-opening parachute operations, allowing offset parachute insertions.

d. Scout swimmer teams whose members are specially trained by the Landing Force Training Command (United States Marine Corps [USMC]) to conduct operations along coastlines, waterways, and inland waters. These teams conduct special insertion, reconnaissance, and security tasks often in support of amphibious operations.

e. Sniper teams, whose members are authorized by the table of organization and equipment (TOE), are used for both conventional and special operations. Along with the two-man sniper teams in the rifle company headquarters, each rifle platoon has two trained snipers.

f Demolition teams whose members are trained in advanced demolition and incendiary techniques and the use of special explosives,

g. Air defense teams trained in the use of man-portable air defense system (MANPADS) weapons. Such teams give the ranger battalion limited self-defense against air attack.

h. Liaison teams that are used by the ranger regimental commander to augment the existing command and control links from the employing unit headquarters to the ranger regiment. These teams are placed where they are most needed and may assist the higher headquarters staff in planning and target selection, as well as performing other liaison duties. See Appendix B for a detailed listing of the personnel and equipment included in the ranger liaison teams.

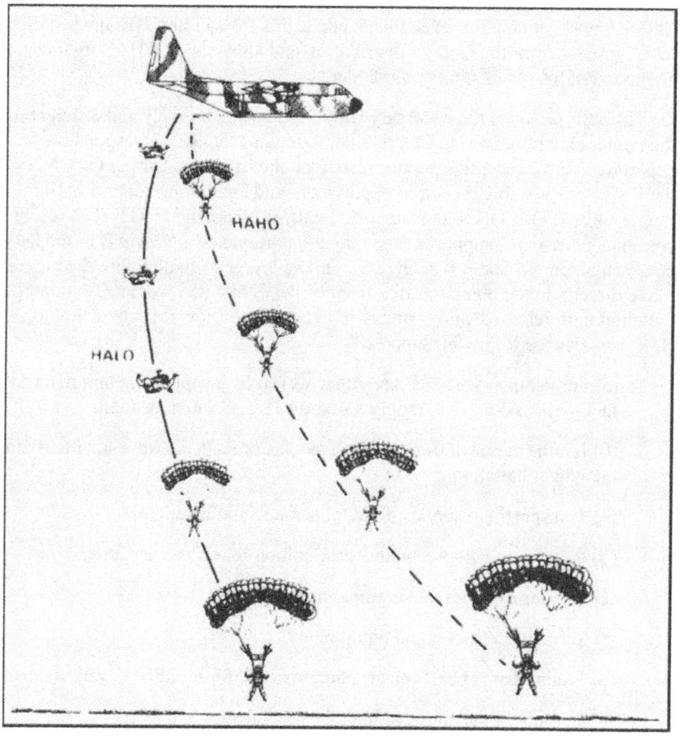

Figure 2-2. HALO and HALO parachute operations.

i. The ranger support element (RSE).

(1) The RSE is a special support element provided by host installation commanders at CONUS posts. It consists of elements and teams from either TOE units stationed at the installation or from the table of distribution and allowances (TDA) organizations under the command of the installation commander.

(2) The RSE provides the necessary support to ensure a timely and efficient deployment of the ranger battalion. During CONUS deployments, the RSE is responsible for both the departure and the reception support provided the battalion. During OCONUS deployments, the RSE is responsible for initial deployment and continuing support until the deploying ranger force is OPCON to the supported unified command or JTF. The responsible theater commander assumes support of the ranger battalion at the REMAB or the ISB if outside continental United States (OCONUS). On deployments that require the ranger battalion to move directly to the objective area from CONUS, the RSE continues support of the ranger battalion until released, and is prepared to accept considerable augmentation assets. The RSE provides the following support:

(a) Furnishing meals that are either served by a supporting unit in an established facility or taken to the ranger battalion at an isolated location.

(b) Issuing supplies from existing stocks, or from contingency items stored for use by the ranger battalion.

(c) Transporting supplies, food, personnel, and equipment.

(d) Palletizing equipment and ammunition for air movement and airdrop.

(e) Rigging vehicles and equipment for airdrop.

(f) Refueling vehicles and equipment.

(g) Issuing barrier and construction material for building of rehearsal sites.

(h) Helping in the construction of rehearsal sites and targets.

(i) Securing, storing, moving, and issuing ammunition and other Class V items.

(j) Maintaining ranger weapons and equipment, to include communications equipment. This is normally done by maintenance contact teams.

(k) Setting up and operating the RSE emergency operations center.

(l) Securing the REMAB if it is in CONUS.

CHAPTER 3
Deployment

Ranger units can quickly deploy by any means of transportation to any location in the world.

3-1. Readiness.

One ranger battalion is always in an advanced readiness condition as the Ranger Ready Force (RRF) available for immediate worldwide deployment. A second battalion is prepared to deploy later and a third battalion to follow. The regimental headquarters maintains command and control, liaison, communications, and reconnaissance elements immediately available for deployment. Higher status of readiness in response to specific world situations can be achieved.

3-2. Ranger ready force.

In order to maintain advanced readiness conditions, ranger battalions are rotated as the RRF.

a. Before taking over the duties of the RRF, battalions must do the following:

(1) Complete preparation for oversea movement (POM) qualification, to include an individual records check and required inoculations.

(2) Test fire all weapons, confirm battlesight settings, and repair or replace faulty weapons.

(3) Assemble and load on pallets the unit basic load (Classes I and V) and other supplies and equipment.

(4) Check unit-derived packing list components for accountability and serviceability.

(5) Prepare special operations troop listings.

b. During an RRF period, ranger units:

(1) Have a positive recall system for all personnel.

(2) Store enough air items for a battalion-size parachute assault.

(3) Conduct no off-post exercises, unless approved by the regimental commander.

c. Each unit keeps an alert roster and a telephone notification system for recalling personnel. The rosters ensure that the duty officer and the unit charge of quarters make a minimum number of calls. Each commander briefs personnel assigned to his unit on the recall system and tests it often.

d. Ranger units can maintain an increased readiness condition for long periods, but they should not be kept in that condition longer than needed. The effort to maintain this level of readiness will degrade the unit's training and effectiveness. If the needs of the operations dictate long periods at the highest readiness condition, provisions must be made for sustainment training. Also, added support will be required from the RSE.

e. The ranger regiment maintains a strong family support system when the unit is training or deployed on combat operations. Each member of the chain of command is responsible for family support.

3-3. N-hour sequence.

The N-hour sequence is a departure time sequence designed to help the ranger force complete all required deployment actions within a certain time. The length of the N-hour sequence depends on the readiness condition of the battalion before deployment. The N-hour sequence begins when the battalion is alerted and ends when the first elements depart. The N-hour sequence does not include travel time to the objective area.

3-4. Intelligence and security.

a. **Essential elements of friendly information.** Items of sensitive information and subject areas, if not secured, can be collectively analyzed by a hostile force. This analysis can reveal the intentions of a unit or compromise a mission. These items of information are designated as essential elements of friendly information (EEFI). They are never transmitted without encryption. The EEFI associated with a ranger unit alert normally are the mission, the use of staging bases, nature of the target, special terrain or weather conditions, and time limits. The ranger force commander must ensure that EEFI are not discussed outside of secure areas or made available to personnel without a security clearance and a need to know.

b. **Security plans.** Security plans for billets, marshalling areas, and out-loading areas must remain flexible to adapt to any situation. Certain items are always addressed in security plans. These are access control and physical security.

(1) **Access control.** Ranger units control access to their respective areas by allowing only authorized personnel to enter or leave. All entrances are secured and an armed guard with the access roster is posted. Access to telephones is restricted.

(2) **Physical security.**

(a) Equipment is open to theft and sabotage during the early phases of deployment. Units must provide guards to secure all unit and personal equipment.

(b) Weapons, ammunition, and sensitive items must be stored in a restricted area and guarded.

(c) Classified documents are open to compromise during deployment. Steps must be set up for maintenance, transportation, turn-in, and destruction of classified material. The most secure facility available must be used.

c. **Rear detachment security.** Security after deployment (maintained by the rear detachment) is as important as security before deployment. No information should be given to the news media from other than official sources. The rear detachment may continue to play a role in the unit cover and deception plan.

d. **Tactical operations center security.** Only personnel listed on TOC access rosters are allowed free entry into the TOC area. Personnel wanting to enter the TOC, who are not listed on the access rosters, are detained until their identity and purpose are proven. Such personnel are escorted at all times while within the TOC area.

e. **Marshalling and out-loading area security.** The ranger unit normally marshalls its personnel and equipment in the unit area, providing its own security force. The USAF security police provide this support at the departure airfield. On Army airfields, Army military police provide security. En route security is provided by the RSE.

f. **Personnel and unit equipment sanitization.**

(1) Before deployment, personnel are relieved of non-mission-essential material having intelligence value. This reduces the intelligence available to hostile forces if they capture a ranger force member.

(2) The ranger force commander sets the degree of sanitization IAW mission requirements.

(3) If strict sanitization is imposed, guidelines are issued as follows:

(a) The only personal items retained are identification card and tags.

(b) Unit insignia and markings are removed from uniforms and equipment.

(c) Personal items, such as wallets and photographs, are secured by the rear detachment commander.

(d) Personnel are briefed on how to compose and retain combat diaries. No personal diaries are allowed.

(4) The commander decides how much US currency is to be taken with the unit. '

g. **Area sanitization.** To avoid the compromise of classified or sensitive mission-related material, searches are conducted before and after leaving the host installation, REMAB, or ISB. These searches are held to locate and remove any material, indicators, litter, or other items that could compromise the ranger unit's mission. Unit commanders ensure that a thorough sanitization of their areas is conducted, to include dismantling rehearsal areas, destroying terrain models, erasing chalkboards, and emptying all classified and nonclassified waste containers. During their searches, commanders must pay close attention to emergency operations centers, tactical operations centers, and marshalling and loading areas. The ranger unit's intelligence officer coordinates support from local counterintelligence elements to search the area again after the ranger unit leaves.

3-5. Operations.

a. **Actions upon receipt of the OPORD.**

(1) Task-organize and equip units for special mission(s) IAW the warning order.

(2) Allocate/reallocate mission-essential equipment.

(3) Distribute augmentee personnel.

(4) Redistribute key personnel,

(5) Prepare a deployment time sequence.

(6) Prepare a movement plan.

(7) Prepare and maintain a current operation estimate.

b. **Advance elements.** A planning element is normally deployed to the REMAB/ISB or JTF headquarters in advance of the main body if the ranger force does not deploy directly to the objective area.

c. **Emergency operations center.** An EOC is setup by the ranger battalion headquarters. The EOC is responsible for relaying all reports to higher headquarters. It remains operational until ordered to cease.

d. **Air movement.** The ranger force always deploys by USAF aircraft. This is normally an administrative air movement to a REMAB or ISB. It may be followed by tactical air deployment from there. For some missions, tactical deployment begins at the unit's home station. Even if the movement is administrative, key personnel are cross-loaded. This is done under the direction of the operations section air officer based on aircraft allocations and load plans.

3-6. Logistics.

The following describes logistics readiness needs and steps for deployment.

a. **Supply.**

(1) **Class I.**

(a) Combat rations are normally palletized when the unit is RRF. These rations are to be used during the tactical operation. If the mission calls for deployment directly into the objective area, the palletized stock of rations is off-loaded in the unit area and issued as needed.

(b) Each ranger deploys with emergency combat rations.

(2) **Classes II and IV.**

(a) **Individual clothing and equipment.**

- Individual A bags are packed IAW a predetermined packing list, tagged, and marked. They are palletized, moved to the departure airfield by the RSE, and loaded IAW the air movement plan.
- Individual B bags are packed, assembled, and forwarded on order by the rear detachment.

- Individual environmental equipment is issued. Equipment shortages are replenished out of stocks maintained by the RSE.

(b) **Air items.**

- While the unit is on RRF, a contingency load of main and reserve parachutes is maintained by the RSE.
- Individual and unit air items are palletized by the RSE and shipped in bulk to the REMAB or ISB.
- For direct-deployment missions, air items and packing material are taken to the unit by the RSE.

(c) **General supplies.** The RSE maintains pallets of common contingency items and special equipment. Pallets are picked up by the out-loading support unit and moved to the unit area for issue or moved with the other pallets to the departure airfield.

(d) **Bulk supplies.** Barrier material is identified, relocated, palletized, and out-loaded IAW the air movement plan.

(3) **Class V.**

(a) The unit basic load of ammunition is secured, palletized, and loaded by the RSE IAW the air movement plan.

(b) For direct deployment missions, selected pallets are taken to the unit area and ammunition is issued. The convoy route is coordinated through the post EOC, and secured by the RSE.

(c) Designated items are rigged by the RSE for airdrop, if enough time exists before movement to the REMAB or ISB begins.

(4) **Class VI.** Personnel deploy with comfort items as needed.

(5) **Class VII.**

(a) All weapons and sensitive items not deployed are assembled and secured by the rear detachment personnel.

(b) Unit vehicles are prepared for airland or airdrop.

(c) Inoperable mission-essential equipment is repaired or replaced.

(d) Special equipment requests are forwarded to the logistics section as they are identified.

(6) **Class VIII.**

(a) Individuals deploy with one set of individual Class VIII items.

(b) Unit medics deploy with modified aidbags.

(c) Controlled medical items (less drugs) are palletized and deployed with section supplies. Controlled drugs are secured and issued to unit aidmen on order.

(d) Medical supply chests are maintained in the unit aid station.

(7) **Class IX.**

(a) Units deploy with initial prescribed load list (PLL) items.

(b) Critical PLL shortages are identified to the RSE for follow-on issue.

b. **Maintenance.**

(1) Upon alert, priority maintenance or replacement of key items of equipment is coordinated by the logistics officer. Only minimum mission-essential equipment is replaced.

(2) Items to be returned from supporting maintenance activities are identified, and pickup is coordinated by the property book officer.

c. **Out-loading coordination and support.**

(1) Upon alert, the RSE provides a liaison officer to the ranger logistics section.

(2) The support platoon leader (SPI,) attends the air movement conference and coordinates the following:

(a) Movement times (for both troops and equipment) to the departure airfield.

(b) Steps for weighing and joint inspection of the pallets and vehicles upon arrival at the departure airfield control group (DACG).

(c) Limitations on the pallet height and weight due to aircraft type, allowable cargo load (ACL), and availability.

(d) Changes in the planned number of pallets due to the mission requirements or airframe availability.

(e) Rigging and loading of the heavy drop and supply bundles.

(3) Empty pallets are placed in position for loading by the RSE in coordination with the SPL.

(a) Companies and staff sections prepare pallets.

(b) For airdrop operations, vehicles are prerigged at the unit location. On order, vehicles are driven to the heavy drop rigging site. The SPL and the officer in charge of the rigger detachment coordinate for a rigging site and the delivery of the required air items to that location. Rigging is done by the RSE, and supervised by a member of the ranger unit. Loading aboard the aircraft is IAW the air movement plan and supervised by RSE and USAF aerial port personnel.

(4) The DACG is provided by the host installation. Upon arrival at the DACG area, unit personnel and equipment are under the control of the DACG. As the deployment begins, a

member of the ranger unit is present at the DACG to ensure an orderly loading of personnel and equipment.

(5) The DACG plans for troop-holding areas in the event of delayed or aborted aircraft.

(6) Unit movement data are maintained by the air officer in the operations section.

3-7. Communications-electronics.

a. **Alert notification communications.** Alert notification of the ranger battalions by the regiment is done by secure communications means. Units in local training areas are informed by radio or through their assigned range field telephone. Units operating out of normal FM radio range during RRF periods set up a high frequency net, or other communications, to their higher headquarters.

b. **Telephone service.** Strict control is maintained over key telephones being used during unit alert. Care is taken to ensure that COMSEC and OPSEC violations do not occur. All telephones are assumed to be monitored; classified, sensitive, or official information is not discussed.

c. **Radio communications.**

(1) **Voice communications.**

(a) A secure radio net is set up early in the deployment. The ranger unit EOC, the installation EOC, and the RSE EOC operate in this net. A liaison officer from the ranger unit is placed at the installation EOC.

(b) The ranger support element liaison officer, attached to the logistics section, operates a secure voice station on a separate logistics net to the RSE EOC and the RSE commander.

(2) **Cryptographic material.** During RRF, COMSEC codes and enabling equipment are kept available to set up the secure voice net.

3-8. Remote marshalling base and intermediate staging base.

a. Ranger units can deploy from their CONUS base directly to the objective area. A more common method would be for the ranger unit to deploy first to either a REMAB or an ISB before insertion into the area of operations. (See Figure 3-1.)

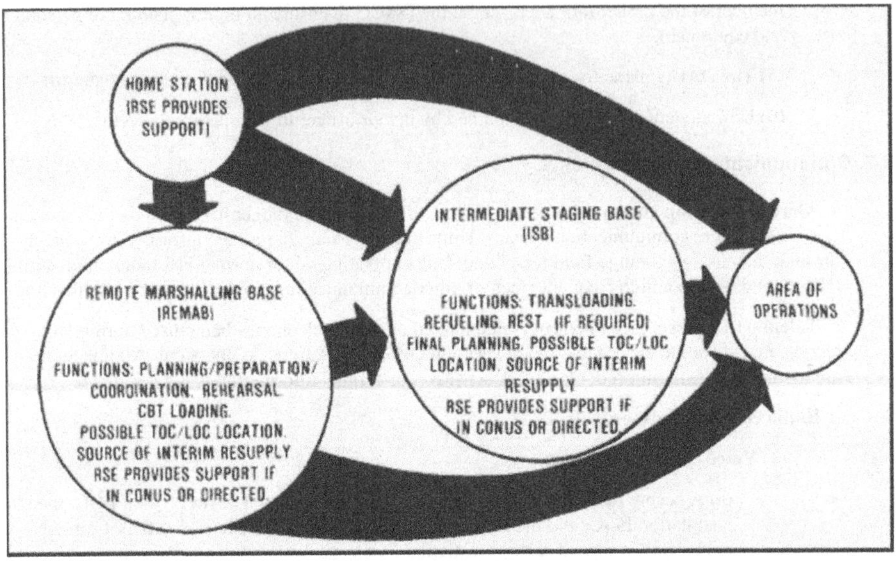

Figure 3-1. Deploying ranger units.

b. The REMAB is a secure base to which the entire ranger unit, including organic and attached support elements, deploys and where mission planning continues. The REMAB is normally within the geographical area encompassed by the command authority of the theater or JTF commander. This ensures that the CSS elements providing support to the ranger unit are operating within their normal area. It prevents or lessens out-of-sector support requirements for CSS elements.

c. The REMAB also provides a secure location for the ranger unit to conduct detailed planning and coordination with the controlling headquarters staff. In the REMAB, the ranger force commander conducts rehearsals, refines and modifies plans, determines PIR, and coordinates with the proper intelligence source to receive that information. It is in the REMAB that specialist augmentees to the ranger force are integrated into the unit, if they have not already joined. Specially trained supporting units, such as aviation and communications elements, also join the ranger force at the REMAB. The REMAB must provide--

(1) Access to the controlling headquarters staff.

(2) Physical security of billeting, planning, maintenance, and communications areas.

(3) Mess, billeting, latrine, and shower facilities for the ranger force and its supporting elements.

(4) Access to a C-141 or C-130 capable airfield, possibly with all-weather operations.

(5) Access to secure communications and processed mission intelligence.

(6) Access to rehearsal areas where sites can be built and live-fire rehearsals can be conducted.

(7) Access to the unit locations of major supporting elements, such as naval landing craft or Army aviation units.

(8) An external security force and an active counterintelligence agency.

(9) Vehicular transport for troop lift, equipment transfer, and administrative use.

(10) Access to maintenance support facilities.

(11) Medical support facilities to augment the ranger force medical personnel.

(12) Covered area to allow for packing of parachutes and rigging of airdrop loads.

d. The ISB is a more austere location. Elements of the ranger force deploy to an ISB to perform final planning, coordination, and task organization. The unit's organization and composition are finalized for movement to the objective area. The ISB is normally not occupied for long periods. Ranger units may deploy to the ISB from the REMAB or CONUS. Deployment to the ISB is more common when terrain or distance precludes insertion to the objective area directly from the REMAB or CONUS. Quick-reaction missions often involve the use of an ISB. Facilities needed at the ISB are limited to the following:

(1) Austere messing arrangements for the combat elements of the ranger force and its attachments.

(2) Austere billets or rest areas.

(3) Water point.

(4) Fuel for aircraft and vehicles.

(5) Areas for test firing of weapons.

(6) Covered and concealed areas for assembly of the ranger force and rigging of parachutes and door bundles.

(7) Austere airfield support facilities, possibly capable of all-weather operations.

(8) Remote location far away from civilians or traffic routes.

(9) Security and counterintelligence elements.

(10) Secure communications.

e. The REMAB and ISB should be located in an area similar in terrain and climate to the objective area. Time spent at the REMAB or ISB lets the ranger force begin acclimatization. Ranger units train yearly in all climates and emphasize physical conditioning. This allows acclimatization quicker than units not as well trained or conditioned. However, ranger units still need an acclimatization period if the terrain and climate of the objective area are different from that of their CONUS base. Not enough acclimatization means that the ranger force could be less efficient, especially when the objective area is hot, dry, or at a great altitude.

CHAPTER 4

Insertion, Extraction, Escape, and Evasion

Insertion is the entrance of a ranger force into a hostile area to execute a particular mission. It is the first critical phase of a ranger operation. These operations often involve deep penetrations of enemy territory by air, water, or land. This chapter emphasizes fundamentals and techniques employed by ranger units when their mission requires insertion and extraction. Extraction is conducted quickly after the mission to avoid casualties.

4-1. Intelligence.

a. Insertion plans are based on timely and accurate intelligence. The headquarters directing the operation must provide current intelligence information to the ranger force. The ranger regiment normally takes action to solicit required information from the controlling headquarters and from other agencies, if security permits. The type of intelligence needed varies depending on the specific mission, but normally includes the following:

(1) Current photographs from aerial or overhead reconnaissance.

(2) Detailed maps.

(3) Data on the population, terrain, and weather from area studies.

(4) Current enemy order of battle, enemy troop locations, and enemy capabilities.

(5) Details on the target area and infiltration routes.

b. To meet its need for current intelligence, the ranger force also relies on aerial reconnaissance and surveillance of the terrain over which it will move. Emphasis is on getting information on the enemy's ability to detect and engage the inserting forces. The location and abilities of air defense radar and weapons systems are critical. Terrain analysis is detailed and focuses on areas suitable for the insertion. Weather information must be detailed and current.

c. In getting information and intelligence data, OPSEC measures are stringent and must be enforced. Coordination with other agencies is kept within security constraints approved by the directing headquarters.

4-2. Deception.

There are always plans to deny the enemy knowledge of the ranger unit's insertion, or to deceive him as to the location or intent of the operation. Feints, false insertions, and other deception operations all add to ranger deception plans. Selection of unexpected means, time, place, and routes of insertion, as well as use of speed and mobility, give less reaction time to the enemy. Planning may include the use of diversionary fires to direct the enemy's attention away from the inserting unit. Deception techniques that may be used include--

a. Multiple airdrops, water landings, or both. This precludes the entire force from becoming engaged at the same time and place, if detected.

b. Dispersion of insertion craft (air or water) both in time and location.

c, Landing a force in an area closer to other potential targets than to the actual targets.

d. Leaks of false information and false messages.

e. False insertions or radio traffic to disguise actual insertions.

f. Diversionary actions such as airstrikes in other than the target areas.

g. Increased reconnaissance flights, either by manned aircraft or remotely piloted vehicles, over false targets.

4-3. Speed and Mobility.

a. Speed of action must be emphasized when the enemy can react quickly to the ranger force presence or when there are limited means of withdrawal. The operation must be conducted swiftly, allowing the ranger force to withdraw before the enemy can react.

b. Major problems facing ranger units are the need for speed and mobility, and the heavy loads carried by soldiers for sustained operations. Ranger units are required to be flexible and to be able to react to any problem within the objective area. However, if equipment is carried to cover every encounter, the load will become too heavy. This will reduce the speed and mobility of the ranger unit. The ranger force commander must always have a balance between the equipment carried and the mobility of his unit. He does this by making three different types of loads: fighting, mission, and existence. He decides what items will make up these loads, where they are carried, and what will be done with them upon contact with the enemy.

(1) The fighting load consists of those items of equipment, weapons, ammunition, food, and water that are common to all soldiers and are necessary for immediate use in combat. These items are normally carried and are retained at all times while in contact with the enemy.

(2) The mission load consists of those items of equipment or ammunition that are required by the mission, but are not normally carried by all rangers in a unit. The load may include spare batteries, demolitions, radios, antitank weapons, or mines. This load is normally spread out among members of the unit and is rotated often. It may be dropped upon enemy contact, but is normally carried to the objective and either used or emplaced there.

(3) The existence load consists of any items designed to provide protection from the elements: sleeping gear, changes of clothes, spare rations, or tentage. This load may or may not be brought into the objective area, depending on the tactical situation. If it is brought, it is normally dropped upon enemy contact or left in an assembly area and retrieved later.

4-4. Stealth.

Methods and steps chosen for insertion must emphasize stealth and try to avoid detection by the enemy. Multiple routes or methods of insertion may be used to preserve the ranger force if some soldiers or units are seen.

4-5. Suppression.

Effort must be made to suppress enemy detection devices, weapons systems, and command and control facilities. This may include electronic countermeasures or direct attack. This helps prevent the enemy from finding and engaging the ranger force en route to the objective area. Deception tactics add to suppression activities. If the ranger force is inserting by infiltration, suppression must not disclose friendly activities.

4-6. Security.

Emphasis on OPSEC prevents compromise of an impending operation. The operation must not be disclosed by overt rehearsals or training activities, the open use or procurement of special items of equipment, or the location of the marshalling area. Several measures can help in maintaining security:

a. Guarded and secure facilities for the headquarters and staff during planning.

b. Issue of operation orders only after units are moved to secure marshalling areas.

c. Alerts, rehearsals, and training to get the local people used to conditions that will exist during actual marshalling.

d. A secure facility set up to isolate units selected for an impending operation.

4-7. Night vision and electronic devices.

a. Rangers use night vision devices to detect and avoid the enemy's forces and his detection devices. These devices also assist in controlling and speeding movement and allow units to traverse almost impassable terrain.

b. The use of electronic homing devices permits precise navigation during reduced visibility. These devices may also be used with visual marking systems.

c. Passive night vision devices are used to assemble rapidly and to reorganize.

d. The use of active night vision devices in assembling, and the degree to which drop zones or assembly areas are "sterilized," depends on the following:

(1) Mission success may depend on not being seen during insertion. When a clandestine insertion is made, the time needed to get to the objective will normally be greater, the use of passive devices will prevail, and the sterilization of the area will be vital. Each ranger must be briefed on sterilization plans and techniques for erasing signs of the insertion.

(2) When detection is likely and the mission depends on getting to the objective before the

enemy can react, speed becomes crucial to the success of the insertion. Insertion may be near or on the objective. Rangers must get in and out quickly. They do minimum sterilization of the area and can use active RSTA devices.

4-8. Rehearsals.

Rehearsals should be close to the actual conditions of insertions and extractions. Security of rehearsals and rehearsal sites is important to the success of the operation. Rehearsals are held on terrain and structures like those in the objective area. If such structures are not available, they are built by the ranger force and the RSE with its supporting engineers.

4-9. Sand tables and terrain models.

Sand tables are useful to acquaint personnel with insertion sites and surrounding terrain. The use of sand tables and terrain models during the issuance of prejump orders and briefings enhances an orderly and speedy tactical assembly.

4-10. Communications.

Reporting during insertion and extraction missions is normally by exception, using an operations schedule.

4-11. Planning insertions.

a. **Centralized planning.** Planning for insertion of ranger units is centralized at the lowest level responsible for the overall operation. This is normally at the joint task force level, with detailed coordination required among all supporting forces. These forces may include other Army elements, USAF, United States Navy (USN), United States Marine Corps (USMC), other US agencies, and allies. Preparation must include detailed contingency planning for use of firepower, shock effect, speed to break contact (if detected), control measures, procedures, and planning. When planning takes a long time, intelligence must be continually updated through all available sources. This includes overhead photography and ground sources to ensure that the initial planning data are still current. Planning stresses the use of deception. Escape and evasion plans, external communication nets, and the location of caches must be known to leaders at all levels.

b. **Reverse planning sequence.** Insertion planning is done in the reverse planning sequence with the ground tactical plan being prepared first. Actions at the objective serve as the basis for the landing and assault plan, the plan for insertion and movement, the loading plan, and the plan for marshalling. All plans are interrelated. Some may be developed at the same time. Consideration of the factors of METT-T guides all planning. Special considerations that apply to these operations are:

(1) **Mission.** The nature of actions at the objective and requirements for special items of equipment and nonorganic personnel impacts on how the unit inserts. When the mission requires rapid deployment, the fastest method of insertion may be needed. In other operations, where success of the mission depends on maintaining secrecy, speed may be less important.

(2) **Enemy situation.** Enemy disposition, strength, and security measures along the route(s) near the objective area affect the method of insertion to be used. Planners must take into consideration whether the enemy:

(a) Uses RSTA detection devices or not.

(b) Has a strong air defense system or possesses air superiority.

(c) Has gaps between his defensive positions.

(3) **Terrain and weather.**

(a) **Topography.** Terrain such as mountains, swamps, or heavy forests favors land infiltration. Deserts, arctic regions, and open prairies favor air or water insertion. Areas not inhabited or with a friendly but scattered population favor land infiltration. A heavily populated area, or one which contains large enemy forces, favors air or water insertion. Hydrography--tides, currents, surf, reefs, and sandbars--must be considered when planning water insertion.

(b) **Weather.** Bad weather with reduced visibility and high winds favors land infiltration, and adversely affects air or water insertion. Limited visibility favors land infiltration; good visibility does not. Adverse weather aerial delivery systems (AWADS) lessens the impact of reduced visibility as a limiting factor on air infiltration. High winds may preclude airborne insertion.

(4) **Troops.** The number of men to be inserted and their level of training may be limiting factors. A need for special skills may call for use of nonranger unit personnel whose physical stamina and ability to perform many infiltration techniques may not equal that of the rangers. Whether supporting forces and special items of equipment are available must also be considered.

(5) **Time.** The main concern is whether the ranger operation is quick-response or deliberate-response. During a quick response mission, planning, preparation, and rehearsal time is short. The timing of the insertion is crucial with respect to weather, the enemy, and if an insertion craft is available.

4-12. Air insertion.

The most rapid form of insertion is by air. Rangers and equipment can be delivered by parachute (static-line or freefall technique), fixed-wing aircraft (airlanded), or helicopter (airlanded, parachute, or rappeling).

a. Suppression of enemy air defense along the insertion corridor may be necessary. The USAF will normally assume the major responsibility for suppressing enemy air defense artillery (ADA) capabilities that may interfere with the inserting forces. This is done by a variety of sophisticated electronic countermeasures applied against enemy equipment and by strikes against known enemy positions. Army fire support elements, special electronic mission aircraft (SEMA), and ground jammers may participate, as well as Navy aircraft, jammers, and gunfire.

(1) A main danger area is the forward edge of the battle area (FEBA) or frontier areas where

the enemy employs most of his sophisticated weapons systems and air defenses. Other dangerous areas are population or troop concentration areas behind enemy lines protected by point air defense systems.

(2) Since the majority of the enemy's detection devices and air defense weapons may be located near the point of entry, fire support, smoke screens (even at night), and suppressive measures may be critical. The use of special items of equipment may be needed to counter the enemy's RSTA effort.

(3) Long-range fires (artillery and naval gunfire) should be planned on known and suspected enemy antiaircraft locations and on prominent terrain features along the route(,s). Once beyond the FEBA, inserting units may be beyond the range of conventional indirect fire. More reliance will then be placed on air assets for fire support and electronic countermeasures.

b. There must be close coordination between the ranger unit and the transporting unit. What the transporting unit does during the insertion has a great impact on the success of the mission. To decrease the chance of detection, maximum use is made of periods of reduced visibility and tactical cover and deception. Drop zones and landing zones are chosen behind tree lines, in small clearings in forests, or on other types of terrain not easily seen.

c. All flight routes over enemy territory should be planned in order to avoid occupied areas and enemy air defenses. Routes should complement the cover and deception plan. They should avoid giving away final ranger objectives.

d. In-flight emergencies, mainly in deep-pedetration flights, should be considered, The ranger force receives a preflight briefing on the route to be flown and is informed of checkpoints along the route. Simple ground assembly plans for contingencies are set up before enplaning. In an emergency, the leader of the unit involved decides whether to continue the mission or abort. This is based on METT-T factors, contingency plans, and distance to the objective as compared to the distance to friendly territory. Contingency provisions for air and water rescue are made.

e. Deciding the DZ location is a joint effort. After receiving input from the air and ground commanders, the joint task force commander makes the final decision. During air assault operations, the ground commander is responsible for the overall process. After reviewing MEYF-T factors, it may be determined that surprise is more important than speed, or vice versa. This may determine the number, size, and locations of DZs and LZs.

f. Planning and preparation continue during the marshalling phase of any operation using air delivery for insertion. A review must be made of items of equipment that are crucial to the ranger force reaching its target area, accomplishing the mission, and returning safely. The need for special supplies and equipment, their transport, the distance to be traversed, the delivery means, and external support are all factors that enhance or reduce the chance for success. Once the means of insertion has been decided, logistics planning continues with emphasis on requesting and coordinating external support. The following must be considered:

(1) Detailed planning for mission logistical needs.

(2) Extra sets of key items of equipment to ensure redundancy and self-sufficiency of

inserting units.

(3) Plans for accompanying, planned, and emergency resupply. The ranger force should prepackage loads to meet specific contingencies. It should consider the use of supplies and equipment already in the operational area.

(4) In quick-response operations, only manportable items of equipment needed to accomplish the mission are carried. Plans and signals must be established for resupply of key items.

(5) Specific plans to evacuate casualties.

(6) All attached personnel and augmentation equipment must be properly prepared for movement.

(7) Provisions must be made for evacuating captured enemy equipment and personnel.

(8) All inserting units should be cross-loaded. Items to be considered in a cross-loading plan are the ground tactical plan, the assembly plan, the configuration of the DZ, the light route and formation of the aircraft, and the number of aircraft available. Cross-loading ensures that key leaders and equipment--for example, door bundles--are evenly distributed throughout the assaulting force. If one or more aircraft abort or are shot down, some key leaders and equipment will still arrive at the DZ. This permits mission accomplishment. Cross-loading aids rapid assembly on the DZ in support of the ground tactical plan. If a ranger is to assemble near the leading edge of a DZ, he is loaded on the aircraft to exit near the front of his stick.

(9) A bump plan is used if an aircraft should abort while still on the ground and there are no spare aircraft. The bump plan ensures that key leaders and equipment are not left behind when the main body takes off. If an aircraft aborts on the ground, key personnel move to another aircraft and replace nonessential personnel. This concept promotes success on the objective.

4-13. Airborne insertion.

a. During the planning phase of an airborne operation, there is no room for error in judgment. It can be planned and conducted well only if there is a constant exchange of information. All personnel involved must be informed of all changes in plans, times, and locations.

b. The reverse planning sequence is the key to the success of an airborne insertion. Most of the planning is based on the ground tactical plan, followed by the air movement plan to support it.

c. The USAF is responsible for flight planning, initial point (IP) selection, and crew procedures throughout the flight. The senior ranger in the aircraft must remain oriented and keep abreast of any last-minute changes. He must coordinate with the aircraft commander.

d. Most airborne insertions are made at either very high altitudes (HALO/HAHO) or very low. For combat operations, drop altitudes may be less than 500 feet above ground level (AGL), depending on the type of parachute used.

e. In airborne insertions, emphasis is on the use of special delivery or navigational techniques since normal insertion is during periods of limited visibility.

(1) **Adverse weather aerial delivery system.**

(a) A self-contained navigational system in some USAF C-130 aircraft is AWADS. The system allows the aircraft crew to fly a specific route to a DZ during darkness or bad weather. The system works with a Doppler radar that determines the ground speed of the aircraft, and an AWADS computer programmed with special flight route information (start point, route, checkpoints, wind speed and direction, and air speed). These two instruments show where the aircraft is in relation to the programmed checkpoints. This eliminates the need for airdrops under visual weather conditions.

(b) Only certain aircraft in formation are equipped with the actual AWADS device. Other aircraft are equipped with station-keeping equipment (SKE). This SKE consists of a computer that tells the aircrew where their aircraft is in relation to other aircraft in the formation.

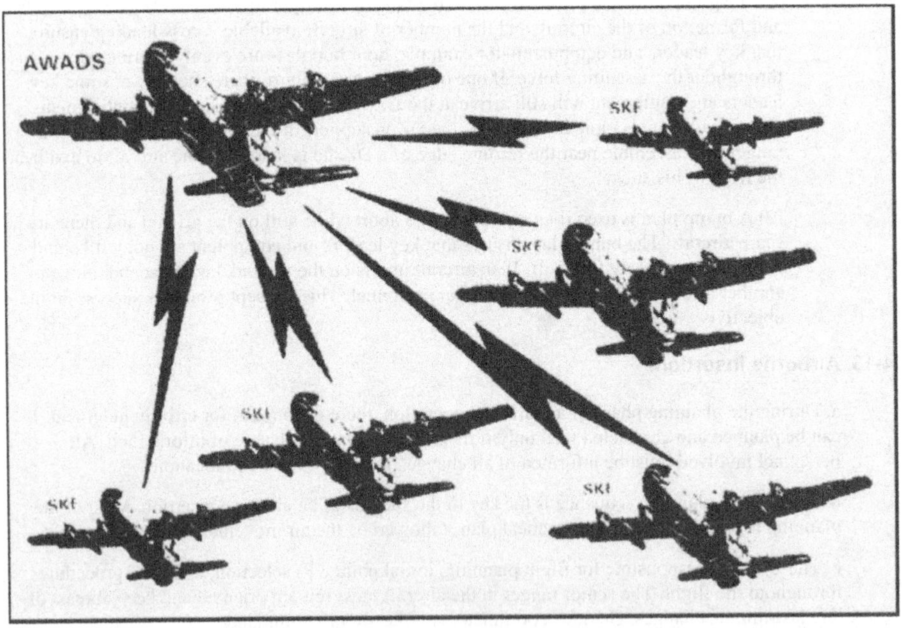

Figure 4-1. AWADS abd SKE aircraft formation.

(c) Airborne insertions using AWADS give the ranger commander the ability to airdrop personnel and equipment during low or zero visibility. This increases the safety of USAF aircraft. It protects them from visually tracked enemy antiaircraft guns and non-all-weather fighters. It also uses adverse weather as a combat force multiplier. Using AWADS, personnel and equipment may be inserted day or night without a pre-positioned USAF combat control team (CCT) or an Army assault team. Time length of the air train is shortened. The air corridor has to be cleared only once. This enhances the element of surprise. The CCTs are brought in with the assault elements. They are able to direct other low altitude parachute extraction systems (LAPES), container delivery systems (CDS), and air-land or heavy drop missions. Use of AWADS enables the ranger commander to conduct rapid, vertical reinforcement of units threatened by an enemy attack during bad weather.

(d) The use of AWADS requires detailed planning, rehearsal, and close coordination among the forces involved. Intense training of both USAF and Army personnel is required.

(e) Personnel must be briefed on terrain through use of photographs, maps, and terrain models. Limited visibility relies mainly on compasses and audible and visual assembly aids.

(f) During reduced visibility, essential heavy-drop serials may precede personnel aircraft. This reduces the time lag between personnel and heavy-drop chalks due to the need to move friendly troops off the DZ. Remaining heavy-drop serials or LAPES can be scheduled following P-hour.

(2) **Reconnaissance teams.**

(a) The operations of the main ranger force may be enhanced by use of reconnaissance teams. These teams are inserted ahead of the main body to provide early warning or to conduct reconnaissance of key objectives. If reconnaissance teams are required at a point along the flight path or near the main DZ, they may be cross-loaded on board any aircraft within the formation. They exit at planned times so as to land as close to their objective as possible. Should such teams be required at a point perpendicular to the axis of the main formation, one of the following is used:

- Another flank aircraft delivers teams to designated points before or with arrival of the main body.
- Selected aircraft deliver the main body, turn, and then deliver the reconnaissance teams.
- The teams are cross-loaded as complete elements on aircraft delivering other units whose flight path crosses the desired locations.

(b) Missions assigned to reconnaissance teams include:

- Providing local security for a USAF CCT, in which case, by joint doctrine, they form the Army assault team (AAT).
- Establishing surveillance or blocking positions on likely avenues of approach into the objective area or DZs.
- Locating and marking targets or assembly areas for the main assault force.

(c) Reconnaissance teams may be inserted by static line or HALO/HAHO operations.

(d) If the reconnaissance element is to locate a specific target or objective, it may be inserted up to 24 hours early. The longer the reconnaissance team remains near the objective area, the greater the danger of it being seen.

(e) In order for the information to be used by the ranger force commander, secure communications and a linkup plan must be prepared and rehearsed.

(3) **Assembly.**

(a) The ranger airborne force can conduct a quick assembly of combat power (weapons, personnel, and command and control elements) on the DZ. Quick assembly and reorganization is crucial because the unit is open to attack. Assembly must be fast and precise. Assembly areas and plans are set forth after considering all the factors of METT-T, mainly the location of the enemy, visibility, terrain, drop information, dispersion pattern, and cross-loading.

(b) The number of assembly areas depends on the location of the DZ, size of the assembly areas, and enemy's detection ability. Unit members move directly to the assembly areas without breaching security. Dissemination of information to all jumpers is the key to a successful assembly plan.

(c) For battalion-size or larger operations, general assembly areas are prescribed in the ranger force OPORD. Company commanders select and designate the exact location of their assembly areas. Although assembly area locations are terrain-dependent, they are generally on the DZ if the drop is at night and off the DZ if the drop is during daylight.

(d) Personnel chosen to carry assembly aids are cross-loaded in the center of their unit sticks. Strong, aggressive leaders should jump first and last in each stick. Upon landing, the leaders roll up their sticks toward the center, locate and move to the proper assembly aid. They may also carry or wear assembly aids. Individual rangers then follow these key personnel. During daylight operations, radiotelephone operators (RATELOS) can display color-coded pennants on their radio antennas to facilitate linkup with leaders. Key personnel may carry and jump with their own radios, and also display color-coded pennants on the radio antennas. Key personnel and RATELOs put their radios into operation before moving to assembly areas. For drops occurring near dawn or dusk, both daylight and night assembly plans are required.

(e) When tactical vehicles or door bundles are dropped, they are normally marked with panels, chemical lights, or strobe lights. The delivery parachutes are often color-coded.

(f) A ranger force should be able to assemble on an unfamiliar DZ, at night, within 45 minutes of the final aircraft pass. Because most ranger operations are time-sensitive, the unit normally moves from the assembly area to the objective when enough of the force has assembled to accomplish the mission. A straggler control point remains at the assembly area under the control of a designated leader. When everyone assembles, or at a specific time, the leader moves to link up with the unit. If available, a radio is left with the straggler control point.

(g) Assembly area security plans are prepared and disseminated. These plans must address hasty defense, sectors of responsibility, observation posts, fire support, and chain of command.

(4) **Contact on the DZ.** During the conduct of parachute insertion, rangers must be ready for enemy engagement at all times. Rangers are open to attack on the DZ before assembling. Just as contingency plans and chance immediate-action drills are developed, briefed, and rehearsed for other chance enemy contacts, rangers must be prepared to execute immediate-action drills for enemy contact on the DZ. The following actions are taken when there is enemy contact on the DZ:

(a) Rangers immediately assemble in fire team-size groups and fight off the DZ to designated assembly areas. Elements attempt to break contact and assemble.

(b) Assembly areas designated by commanders in the mission OPORD serve as primary rally points.

(c) Alternate assembly areas (normally 2 to 5 kilometers from the DZ) are designated in the mission operation order (OPORD). They are used by elements that cannot link up or assemble in the primary assembly area. These elements continue with the mission once assembled.

(d) During limited visibility, personnel should move directly and quickly off the DZ. They must avoid mistaking other friendly elements for enemy, which could cause firefights among friendly forces.

(5) **Reorganization.**

(a) The unit reorganizes according to plans, using designated assembly areas, aids, and identification markings for personnel and equipment. Replacements are named for key leaders or personnel who are lost or injured.

(b) Security teams assemble and move to their positions. Remaining units move quickly to their assembly areas, carrying supplies and equipment.

(c) Subordinate leaders keep the commander informed during assembly and reorganization.

(d) Runners may be used to report assembly status. This maintains radio silence. Designated personnel quickly recover supplies and equipment and move to the assembly areas.

(e) Reorganization is complete when all units have assembled or have been accounted for, and control has been established.

(6) **Pop-up technique.** This involves close terrain flying until the DZ is reached, then the aircraft quickly climbs to jump altitude. Special navigational equipment and aircrew training

are needed but may not always be available.

(7) **HALO/HAHO.** This involves high-altitude, low-opening and high-altitude, high-opening jumps with high performance parachutes designed to maneuver the jumpers to a specific point on the ground. During these operations, midair assembly procedures may be used.

(8) **Arctic.** Airborne insertion in extremely cold conditions is difficult. Times allocated for all parachute and equipment rigging must be increased. Rigging of individual equipment must be IAW FM 57-220. Ahkios can be rigged, packed inside a double A22 container, and dropped in platoon bundles. Because of the increased amount of individual equipment needed, allowable cargo loads for USAF aircraft are reduced.

(9) **Water.** Because of a lack of suitable land DZs, or to avoid detection, rangers may conduct airborne operations onto water DZs close to land. Procedures and techniques are similar to land DZ operations.

(10) **In-flight rigging.** On long-range operations involving extended flight time to the objective area, personnel may rig in flight. In-flight rigging requires detailed planning, coordination, of all jumpers. A rehearsal of these procedures should be conducted before deployment. Aircraft loads must be reduced to carry out in-flight rigging procedures. (See FM 57-220.)

4-14. Air-land insertion (fixed-wing).

a. **Planning factors.** Most of the planning factors that apply to airborne insertion operations also apply to air-land insertion operations. Most USAF aircraft cannot conduct air-land operations in total instrument weather conditions. The AWADS equipment is accurate enough for airdrop navigation, not air-land. Air-land operations are conducted, however, under conditions of extremely reduced visibility. Specially trained USAF crews, using sophisticated navigational techniques, can locate and land on small airfields in bad weather. As in airborne operations, ranger reconnaissance teams, as well as specially trained CCT elements, can jump before the air-land aircraft arrive. They emplace navigational and location markers to assist the pilots. Rangers may rely solely on air-land operations for insertion. This method does not provide for delivery of the force in the shortest time possible. Air-land insertions are normally in concert with an insertion of airborne elements.

b. **Time.** Since turnaround time is crucial, loads in aircraft should be unchained as the aircraft taxies. This also applies to personnel and seat belts. Aircraft should stop as close as possible to the assembly areas where personnel and equipment are off-loaded.

c. **Method.** Air-land insertions normally begin after a ranger Army assault team (AAT) and a USAF CCT have parachuted into the initial objective area and conducted reconnaissance, clearing, marking, and security operations. The ranger ground force commander is the arrival airfield commander. He has control over all arriving, taxiing, and departing aircraft (fixed-wing or rotary) through the CCT, which locates near him. The ranger force command and control element is usually on the first aircraft in and the last aircraft out.

d. **Security.** Air-land operations concentrate large numbers of men and equipment, which presents an open target to the enemy. The ranger ground force commander must ensure rapid off-loading, assembly, dispersion, and movement toward the objective.

e. **Communications.** The ranger ground force commander must have secure radio communications with the ground force, CCT, fire support elements, and higher headquarters. Wire should be laid to connect ground command and control locations.

f. **Operation duration.** Depending on the factors of METT-T, the aircraft may remain on the ground for extraction while the mission is being accomplished or take off at once. They may then orbit, pending recall, or return at a prearranged time.

g. **Extraction.** Extraction of the air-landed ranger force is normally planned and accomplished like an insertion, only in reverse order. A simple, yet effective, way must be found to see that no ranger personnel are left in the objective area.

4-15. Air assault insertion.

Ranger air assault (air-land, rappel, or combination) insertions normally take place during limited visibility. They quickly place the ranger force on, or close to, the objective. The ranger force (depending on the mission) normally plans on fighting when it hits the ground, completing the mission, and then being extracted. Air assault insertions need detailed planning, surprise, flexibility, speed, shock effect, concentration of combat power, and precise timing.

a. **Ground tactical plan.** The reverse planning process is important. The ground tactical plan, developed from the mission assessment, is the first planning area. All other planning begins from this point. The ranger commander and his staff decide stated and implied tasks, and request the proper helicopters for the mission.

b. **Fire support plan.** The fire support available to the ranger commander may be artillery, naval gunfire, mortars, attack helicopters, and USAF or USN tactical aircraft. Because of the timing of ranger air assault operations, preparatory fires are often not used. If they are used, enemy ADA sites normally have priority.

c. **Landing plan.** This plan puts the ranger force into the objective area at the time, place, and proper sequence to support the ground tactical plan. The ranger commander selects primary and alternate LZs and the priority for landing. The exact location of an LZ is determined after a detailed analysis of METT-T. Rehearsals are important. The ranger force must quickly assemble, reorganize, and accomplish the mission. The senior ranger on each aircraft ensures that the personnel on his aircraft offload at the correct LZ.

d. **Air movement plan.** The air movement plan provides for the coordinated movement of the ranger force into the zone of action in support of the 'landing plan. Flight routes, air movement tables, flight formation, altitude, and air speed should be considered. The ranger commander must have secure in-flight communications with his subordinate commanders.

e. **Loading plan.** This plan supports the air movement plan and provides for the pickup zone (PZ) setup and control. It sets priorities for the commander for loading personnel, equipment, supplies, and cross-loading. Air-loading table, aircraft loading formations, bump plan, and rehearsals should be considered.

f. **Staging plan.**

(1) This plan includes PZ selection and operation. The selection of PZs and LZs needs planning and coordination for best use of air assault assets. Site selection in or near the objective must be coordinated in person between the supported ranger commander and the aviation commander. The tactical situation is the key factor in PZ and LZ selection. Other factors are size of landing zones, surface conditions, ground slopes, approach and departure directions, prevailing winds, obstacles, communications, cross-loading, aircraft command and control, PZ and LZ markings, and rehearsals.

(2) The PZs are set up and secured about 15 minutes before touchdown of first aircraft. The fire support plan must support the other plans. All supporting fires must be coordinated with the air mission commander (AMC). Because of the depth of most objective locations, USAF tactical aircraft will be the main fire support method. Orbit locations for fire support aircraft must be planned.

g. **Other planning considerations.** These include escape and evasion, actions at the last LZ, assembly plan, downed aircraft procedures, control measures, forward area rearm and refuel point (FARP), weather delays, and deception plans.

4-16. Amphibious insertion.

Insertion by water may be by SCUBA, surface swimming, small boats, submarines, surface craft, helocasting, or a combination of these methods.

a. **Planning.** Planning for waterborne operations must be thorough. The plan is decided upon between the ranger commander, the transporting unit, and the joint headquarters directing the operation. Plans must be made for all types of enemy action and weather while on board the transporting craft. The transporting unit is given information on a need-to-know basis. Information that could endanger the operation may be withheld until the mission is under way. Initial planning includes:

(1) Time schedule of all events, from the beginning until the end of the joint operation, to be used as a planning guide. Timing for each event must be exact; the success of the operation depends on it. (See Figure 4-2)

Figure 4-2. Waterborne insertion.

(2) Embarkation site, where troops and equipment are loaded on the transporting craft.

(3) Drop site, where the ranger force will depart the primary craft and load into smaller boats.

(4) Handing site, where the unit will beach its boats, or land directly from amphibious craft.

(5) Loading, when loads are lashed and waterproofed IAW unit standing operating procedures (SOPs). Inspections are important. A bump plan for key personnel and equipment must be developed.

(6) Cross-loading, which will help preserve the ranger force. location of leaders, crew-served weapons, and needed equipment are stressed.

b. **Landing site.**

(1) **Beaches.** Beaches are checked to find one that allows an undetected approach. Landing sites that cannot be approached from different directions are avoided. Stretches of defended or patrolled coastline should be avoided. If sand beaches are used, tracks and other signs of

operation that may compromise the mission should be erased. Rural, isolated areas are best. The area behind the landing site should provide an assembly area and concealed routes for exit.

(2) **Barriers.** Reconnaissance may find that good landing sites are not suitable because of barriers between the site and the objective. Such barriers may be patrolled roads, large areas giving little or no concealment, enemy obstacles or defensive positions, areas populated with enemy sympathizers, or an area used for enemy troop maneuvers and training. Unless a means of avoiding such barriers is devised, landing sites should be limited to those without barriers.

(3) **Other factors.** Beach selection should also consider:

(a) The scheme of maneuver.

(b) Surprise.

(c) Enemy dispositions.

(d) Distance to the target.

(e) Landing and exit sites.

(f) Trafficability of the beach.

(g) Cover and concealment.

c. **Routes.** The route the transporting craft takes to the drop site should be planned to deceive the enemy. A major change in the route immediately after the unit's debarkation could compromise the mission. The route should be the same as a route that would be used in some other naval operation--for example, minelaying or sweeping, or a naval patrol. An alternate route must be planned.

d. **Tactical deception.** There should be other plans to deny the enemy knowledge of the insertion besides the water approach route plan. Plans may include electronic countermeasures (ECM), diversions, feints, ruses, or demonstrations.

e. **Actions at the drop site.** Some operations may permit landing directly from the transporting craft. If not, primary and alternate drop sites must be set. The drop site should be at least 1,500 meters offshore to cover any noise in the loading and launching. If the enemy has a surface radar capability, a drop site should be chosen to avoid detection. The drop site may have to be several miles offshore or may need the use of ECM. The manner in which men, small boats, and equipment are loaded must promote fast action at the drop site.

f. **Navigation.**

(1) Ship-to-shore navigation (to the landing site) may be done by dead reckoning. The course may also be maintained by compass navigation, reference to a shoreline silhouette, radar, or prearranged signals from the shore.

(2) Boat handlers, navigators, and other experts are relied on in various phases of the

transport operation.

(3) Lead reconnaissance teams or scout swimmer teams may be used to plant radar, radio, or infrared buoys, or distinct, shielded lights. Boats may be equipped with radar reflectors or beacons to permit accurate tracking.

g. **Actions on the beach.**

(1) Boats hold offshore while ranger scout swimmers move to a covered and concealed position. After reaching the beach, they make a short listening halt. They check the beach and the area immediately beyond the beach for enemy activity. Swimmers signal the force "all clear," mark the landing site, and provide local security.

(2) When the boats land, chosen personnel move directly into covered and concealed security positions. They prepare to defend the landing site.

(3) If the ranger unit is not carrying equipment, boats may be moved off the beach and buried or camouflaged. If heavy equipment makes it hard to move the boats, chosen personnel start unloading at once. The remainder of the force secures the landing site. Each boatload moves into position under the control of a designated leader. Teams are positioned with the best cover, concealment, and fields of fire available.

(4) The ranger force leader sets up a command post from which he can control the operation. He may establish secure communication with his headquarters.

(5) Boats may be deflated and buried or camouflaged near the landing site or away from it. If the boats are to be hidden near the landing site, a team may be chosen to dig holes or cut brush for camouflage along with the unloading of the boats. A small team may be chosen to sweep the beach to erase tracks and drag marks.

(6) The ranger force should be assembled and prepared to move within minutes after the last boat lands. Every ranger must be briefed on the unit's plan should enemy contact be made upon landing.

h. **Insertion by air from ship.**

(1) Helicopters from a ship can launch the ranger force from over the horizon and move at a low level to the LZ.

(2) Helicopters may be vectored from ships to a landing zone. The aircraft may return to the ship while the ranger force conducts the mission, they may orbit offshore, or they may stay at the LZ. Other aspects of landing and assembling are the same as discussed for helicopter insertions.

i. **Helocasting.** This form of insertion combines helicopters and small boats in the same operation. It is planned and conducted the same as air assault operations, except that the LZ is in the water. While helicopters are moving at low levels and low speeds, rangers drop small boats and jump into the water. They then assemble, climb into the boats, and move to shore. Ranger elements are usually assembled and loaded into the small boats within 20 minutes of the last ranger entering the water. This type of operation is normally conducted during limited visibility.

j. **Fire support.** Fire support may be ship-to-air, ship-to-shore, air-to-ground, or ground-to-ground. Specific missions include:

(1) Suppressing enemy air defense.

(2) Neutralizing enemy field artillery within range of boat or helicopter routes.

(3) Employing on-call fires against shoreline targets.

(4) Destroying enemy aircraft.

k. **Contingency planning.** The following must be planned for:

(1) Actions on enemy contact.

(a) En route.

(b) Helocast site.

(c) Flares.

(d) Aerial attack.

(e) Small arms fire.

(f) Indirect fire.

(2) Bump plan with priority to key leaders, crew-served weapons, and radios.

(3) Downed craft procedures.

(4) Escape and evasion.

(5) High surf.

(6) Adverse weather.

(7) Separation.

l. **Rehearsals.** The ranger force must rehearse all aspects of the amphibious insertion. This includes boat launching, paddling, beaching and assembly, boat commands, and capsize drills.

4-17. Land infiltration.

Land infiltration to the objective may be the best way to complete a mission. This is normal when the enemy has total air control or effective air defenses in the operational area to be penetrated, or when the objective is close to friendly territory. Rangers can infiltrate over any type of terrain, in any climate. However, thick forests, swamps, and broken or steep terrain offer the best chance for success.

a. **Planning.** There are many methods of infiltrating by land, and each has its good and bad points. Plans are designed to get the infiltrating force to the objective area without being seen. Plans should include:

(1) Selecting concealed routes for surprise and protection. Primary and alternate routes are based on detailed map and aerial photo study, ground reconnaissance, and data from agent reports.

(2) Avoiding obstacles, populated areas, silhouetting (when forced to cross obstacles or ridge lines), enemy positions, main avenues of approach, and movement along often used routes and trails.

(3) Selecting the time of infiltration to make use of reduced visibility, such as darkness, rain, snow, and fog, and periods of reduced enemy alertness, such as early morning, holidays, and times following protracted movement or combat. The time of the infiltration is important during critical phases (crossing borders or passing through enemy troop concentrations or populated areas).

(4) Using speed and stealth. All rangers must know routes, linkup points, alternate linkup points, time schedules, danger areas, and actions on enemy contact and at linkup points.

(5) Centralizing planning and decentralizing execution. Centralized coordination must ensure that units are acting IAW cover and deception plans. Control should be given to the leaders of the infiltrating groups. Within their designated zones, these leaders select their own routes and formations. Leaders control the rate of advance. When the enemy has electronic detection devices, radio communications must be kept to a minimum. Key leaders normally go with the group that is using the fastest lane or the lane least likely to be detected.

(6) Seeding infiltration routes to the assembly area with unattended ground sensors. These sensors can help the infiltrating force by reducing the chance of surprise contact during the infiltration, and by helping in controlling the forward progress of the units.

(7) Planning for fire support, which must include coverage of routes, danger areas, assembly areas, objectives, and PZs.

(8) Rehearsing actions at the objective, actions on enemy contact, actions at danger areas, and movement techniques.

b. **Movement.** Land infiltration is conducted using one of three techniques:

(1) **Movement in small groups along one axis.** All members of the force use the best route. Small groups are harder to detect, easier to control, and do not compromise the total force if detected. This technique may take too long and it needs an assembly area or linkup point. If the lead group is detected, other groups may be ambushed.

(2) **Movement in one group.** This technique has no reassembly problem. The same route is used by all, making navigation easier. A large force can fight out of a dangerous situation easier than a small one. It has no coordination problems between other infiltrating units. This form of movement is easier for the enemy to detect. If the force is detected, the mission may be endangered.

(3) **Movement in small groups along several routes at the same time.** This technique avoids putting the total force in danger, is less likely to be seen, forces the enemy to react in

many locations, and makes it harder for the enemy to determine the size of the force or its mission. Some groups travel over poorer routes and have problems of assembly, control, and support; the capture of personnel may endanger the mission. Regardless of which technique is used, ranger units normally move to reduce the chance of enemy contact.

c. **Assembly area.** Following movement into the general area of operations, infiltrating units converge on the assembly area or linkup point. Assembly may begin at once after the passage of lines, or it may take place at any time before reaching the objective. The infiltrating force must remain undetected before assembly. If an infiltrating unit is seen at this time, the enemy may learn the location of the assembly area, and the objective. It may take action to destroy the infiltrating force.

d. **Linkup point(s).**

(1) Selection characteristics include terrain that--

(a) Is free of enemy troops.

(b) Offers cover and concealment from enemy observation and patrols.

(c) Does not restrict movement of friendly units.

(d) Accommodates the unit.

(e) Is easily identified.

(f) Offers escape routes to alternate linkup points.

(g) Offers concealment from air attack.

(2) Action sequence for occupying the linkup point.

(a) Halt short of the point.

(b) Establish local security.

(c) Conduct a reconnaissance of the linkup point.

(d) Occupy the linkup point.

(e) Emplace observation posts and listening posts.

(f) Link up with other groups.

(g) Conduct final preparations.

(h) Continue the mission.

(3) Security may be enhanced with the aid of devices such as night vision goggles and night sights. Unattended ground sensors may be placed on likely avenues of approach into the linkup point. Normally, only passive observation devices are used to assist in avoiding detection.

(4) Plans must set a time schedule for actions at the linkup point. All unit members must move in quickly, consistent with security, so that the unit may proceed to the objective rally point (ORP) and on to the objective.

4-18. Stay-behind technique.

A ranger force can be positioned behind enemy lines by employing the stay-behind technique. In this method, a ranger force allows itself to be bypassed as the enemy advances. This is done to allow the ranger force to perform a specific mission. Stay-behind operations may require the concealment, or cache, of supplies before the enemy bypass. Rangers may need to use captured enemy supplies. The location of the assembly area and the plan for extraction are important parts of stay-behind plans.

4-19. Extraction.

The ranger force is normally extracted after the mission is completed. The enemy may also force the unit to abort its mission and to call for extraction. Contingency plans must be made to extract a force under enemy pressure, as well as one that is not in contact. Primary and alternate extraction sites are planned and coordinated with all ,porting forces. Ranger operations are normally conducted against deep targets. Extraction may begin with land movement but normally it includes the use of boats or aircraft. Plans may include primary I alternate means of extraction.

a. **Terrain.**

(1) The terrain is important in choosing the point of extraction - for example, whether or not the extraction site offers good cover from enemy direct-fire weapons and positions from which the extracting force can suppress the enemy. Tidal conditions and PZ size must be considered.

(2) Unlikely locations such as swamps, jungles, or mountain areas are often used for extraction. Specially trained air and naval crews, using the latest extraction devices, make such terrain useable.

b. **Enemy.**

(1) Extraction is best conducted without enemy pressure. However, such pressure could develop during the operation. Detailed plans are made for emergency or contingency extractions forced by the enemy as well as clandestine extractions not under enemy pressure. If under pressure, units will have to use planned fire support to suppress enemy weapons. At the beginning of the extraction with security posted to provide early warning, the unit may start to thin out the same as with a night withdrawal. Ground security may be relied on until the last unit withdraws under cover of suppressive fires.

(2) If enemy contact is made during an extraction, the commander must decide whether to reinsert units or break contact and move to another extraction point. Fire support and suppression are important and include suppressing enemy air defense by deception, ruses, or decoys.

(3) The ranger unit may have special weapons systems or equipment to aid in the extraction. Shoulder-fired antiaircraft missiles may be used to defend against enemy aircraft threatening

the extraction.

c. **Extraction by air or water.** Extraction by air or water is preferred when the resources are available and their use will not endanger the mission. Other factors that favor their use are when--

(1) Long distances must be covered.

(2) Time of return is essential.

(3) The enemy does not have air and naval superiority.

(4) Heavily populated hostile areas block land movement.

(5) Ranger forces are burdened with prisoners of war,casualties, or critical items of equipment.

d. **Extraction by land exfiltration.** Stealth and evasiveness are key elements of exfiltration. This method of extraction is favored when--

(1) The enemy has air superiority.

(2) The enemy is able to prohibit air or water extraction.

(3) The distance to friendly lines is short.

(4) The terrain provides cover and concealment (for movement on foot) and limits enemy mobile units.

(a) Multiple routes of exfiltration may be used to prevent destruction of the force if soldiers or units are detected. Captured enemy vehicles and equipment may be used to assist in the exfiltration.

(b) The ranger force may exfiltrate in one body or in small groups. Exfiltrating in small groups avoids the time delay of assembly.

(5) The exfiltrating force is lightly equipped and not burdened with captured personnel or material.

(6) The exfiltration route passes through an area occupied by friendly civilians or guerrilla forces who can assist the movement.

(7) Areas along exfiltration routes are not inhabited.

(8) The enemy force is dispersed or is not able to concentrate against the exfiltrating force.

4-20. Escape and evasion.

a. A vital part of all premission planning is the development of en-route and post mission E&E plans. Such plans enhance survival of the ranger force, and the transporting air or boat crews. The ranger force commander is responsible for--

(1) Helping in the development of an E&E plan in coordination with all supporting agencies.

(2) Ensuring that all members of the ranger force and the supporting boat or air crew are briefed on the E&E plan.

b. Each E&E plan is unique. Each situation has unique problems associated with E&E. The E&E plan devised by the ranger force commander must address these problems, while profiting from the individual abilities and training of the rangers and their supporting air and boat crews. The following considerations apply to E&E plans devised during ranger operations:

(1) The plans enhance survival of personnel who can no longer accomplish their assigned mission. The senior ranger in a group must decide if there are any remaining missions that the group can accomplish. If not, then he must try to evade and escape enemy capture if unable to link up and be extracted with the rest of the ranger force.

(2) Because of the depth of penetration behind enemy lines, most successful E&E plans may involve either air or water movement away from enemy-held territory. The air and boat crews supporting ranger operations are highly skilled and dedicated. If there is any chance of a successful rescue, these forces continue to try to retrieve rangers escaping from an enemy-held area.

(3) Rangers possess stamina and endurance. They can move a great distance (especially at night) and over rugged terrain to reach an area where they can rendezvous with search-and-rescue (SAR) aircraft or boats. Rangers are also resourceful and highly trained in communication, stealth, and camouflage. This, coupled with their self-discipline, means that successful E&E can be conducted over a great distance, through heavy concentrations of enemy, and over a long period.

(4) Escape and evasion plans for ranger elements should include avoiding contact with the local civilians. However, the aid of friendly insurgent forces may be enlisted. The ranger E&E plan may also include the use of other special operations forces E&E networks that are in place behind enemy lines. These nets must not be compromised by the volume of evaders.

c. Escape and evasion plans should be based on three phases:

PHASE ONE: That portion of the insertion following entry into the hostile zone where personnel survival and recovery is the only thing considered. Should an aircraft or boat carrying rangers be disabled during phase one, the ranking survivor assumes command. He must consider E&E planning and the experience and skills of other survivors. He should be concerned with the security of the mission, caring for wounded, and moving away from enemy activity while contacting SAR elements for recovery.

PHASE TWO: That portion of the insertion that is near the objective and could permit the ranger element to successfully pursue its mission. Although part of the E&E plan, actions by the ranger force during this phase become a forced land infiltration rather than an E&E. The senior surviving ranger must decide if enough personnel and equipment can be moved to the objective area in time to accomplish the mission. If so, he must proceed, taking along any surviving air or boat crew members. If not, he may proceed directly to the extraction point and wait, or he may contact SAR elements for recovery.

PHASE THREE: That portion of the operation after mission accomplishment. It is a hard time to escape and evade since the enemy is alert and reacting to the ranger forces. The confusion caused by the violence of the attack, coupled with the rangers' ability to move quickly cross-country away from the objective area, offers the best chance of a successful escape and evasion. The surviving ranger elements that cannot be extracted can do one of the following:

(1) Move over land at once to a PZ or beach for extraction by SAR elements.

(2) Move to an area where reentry into friendly lines can be made.

(3) Move to a neutral or nonaligned country where either internment or return to friendly control can be effected.

d. Any plan for E&E should provide for the issue to each ranger of an area map, a covert means of recognition, and a small compass. An emergency ration may also be included and any items needed for the climate in that objective area. The ranger force commander must balance the need for other survival equipment against the added weight and bulk involved.

CHAPTER 5

Strike Operations

The ranger regiment conducts strike operations in support of the Air Land Battle. Ranger strike operations are integrated into an overall plan designed to destroy, delay, and disorganize the enemy. They may also cause him to divert a major part of his combat forces and power to rear area security. Strike operations also help create an environment in which other special operations forces can operate. Strike operations can complement air power in disrupting enemy forces in depth. They allow the theater commander to regain the initiative and lead to decisive action.

5-1. Application.

a. The ranger regiment can conduct strike operations either as deliberate-response or quick-reaction missions. It can be inserted into the area of operations by land, sea, or air.

b. Strike operations can be conducted throughout the depth of the battlefield in support of conventional or unconventional warfare. Multiple strike operations can have a synergistic effect and can create situations that help friendly forces seize and maintain the initiative.

c. Strike operations include raids, personnel and equipment recovery, and interdiction of lines of communication. They are best accomplished by ranger units who use conventional techniques and special skills. Strike operations vary in size. When well augmented and supported, the ranger regiment can conduct strike operations deep in enemy territory using up to three battalions. Strike operations normally are limited in time and may be conducted unilaterally or in conjunction with other special operations forces or allied forces. They are normally ended by the extraction or exfiltration of the ranger force.

5-2. Raid operations.

a. A raid is a strike operation conducted behind enemy lines against strategic objectives, targets of high tactical value, time-sensitive targets, or key personnel and facilities in enemy rear areas. It normally involves a swift penetration of hostile territory to confuse the enemy or destroy his installations. Ranger units can conduct deep penetration raids when given the appropriate assets for insertion.

b. A strategic raid by the ranger regiment would be directed by the NCA and conducted under the OPCOM of a unified or JTF commander. An operational raid would be directed by the unified or JTF commander and conducted by that command directly or under the OPCOM of a corps commander. Typical targets would include:

(1) Command, control, communications, and intelligence centers at front level or combined arms army level.

(2) Nuclear and chemical weapons storage sites and delivery means.

(3) Key logistical centers, warehouses, ammunition complexes, or fuel pumping centers supplying logistical support to a specific front.

(4) Air defense and air traffic control integrating centers and air defense weapons sites.

(5) Key power generating and transmitting facilities, hydroelectric dams, and irrigation systems.

(6) Key ports or rail complexes.

(7) Key installations or facilities, such as airstrips, buildings, bridges, dams, or tunnels.

c. The size of the ranger raid force is determined by a detailed analysis of METT-T. The unit committed to the ground phase of the raid mission is kept as small as possible. It may be as small as a ranger company or it could include all the battalions of the ranger regiment. The force must be tailored to complete the mission quickly, violently, and with few casualties. The size of the force may also be affected by the augmentation of specially trained personnel needed for that mission.

d. Raids are normally conducted in five phases. (See Figure 5-1).

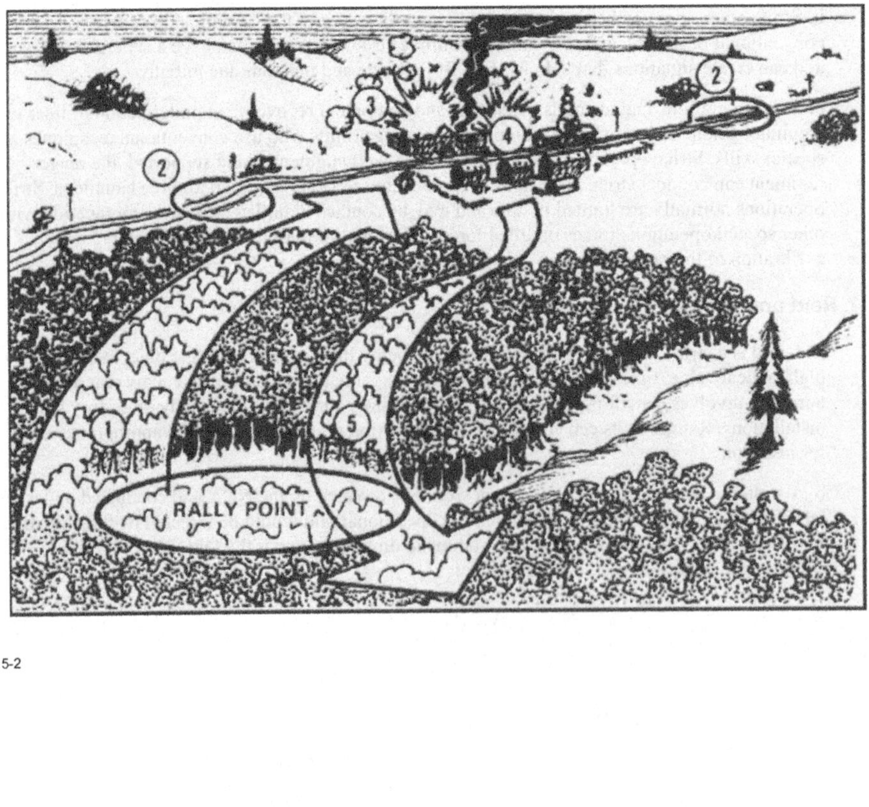

Figure 5-1. Five phases of a raid.

PHASE 1. The ranger force inserts or infiltrates into the objective area.

PHASE 2. The objective area is then sealed off from outside support or reinforcement, to include the enemy air threat.

PHASE 3. Any enemy force at or near the objective is overcome by surprise and violent attack, using all available firepower for shock effect.

PHASE 4. The mission is accomplished quickly before any surviving enemy can recover or be reinforced.

PHASE 5. The ranger force quickly withdraws from the objective area and is extracted.

e. This paragraph discusses the raid after insertion/infiltration. All considerations discussed in Chapter 4 must be applied in order for the unit to get deep behind enemy lines. Upon receipt of a raid mission, the following basics are emphasized:

(1) **Maximum use of intelligence information.** The gathering and dissemination of information must be continuous and provided to the raid force even while en route to the target area. To ensure mission accomplishment, the ranger unit must be kept informed of the latest enemy developments in the objective area to prevent being surprised. Only then can the full combat power of the rangers be concentrated at the decisive time and place.

(2) **Plan development.** The reverse planning sequence and the planning steps set forth in FM 101-5 are followed in the plan development.

(3) **Coordination.** Coordination is through the operational headquarters with friendly units (such as higher and supporting Army or joint headquarters, intelligence agencies) for fire, RSTA, special equipment, and personnel and logistical support.

(4) **Rehearsals.** Rehearsals help validate all aspects of planning for the raid and ensure precision in execution-They allow changes to be made in the plan before it is carried out.

f. A successful raid is ensured by--

(1) Launching the raid at an unexpected time or place by taking advantage of darkness and other periods of limited visibility, and moving over terrain that the enemy may consider impassable.

(2) Avoiding detection through proper movement techniques and skillful camouflage and concealment to include taking advantage of natural cover of the terrain, and using sophisticated equipment to detect and avoid enemy forces.

(3) Timing the operation as close as possible.

(4) Using all available support, both organic and nonorganic, to include use of special weapons such as Air Force "smart bombs" and artillery cannon-launched guided projectiles, with the ranger unit using laser target designators.

(5) Performing quick, violent, precise, and audacious actions that focus full combat power at the decisive time and place.

(6) Disengaging quickly upon mission completion.

(7) Withdrawing swiftly using planned routes and including a deception plan.

g. Four functions are normally performed by ranger tactical formations when conducting a raid. Each subelement is organized and equipped to do a specific part of the overall mission. Depending upon the specific mission, nature of the target, enemy situation, and terrain, these four functions are as follows:

(1) The command group controls movement to and actions at the objective. This unit normally consists of the ground commander, other subordinate leaders in the raid organization, and communications to support these leaders.

(2) The security element, whose organization is determined by the mission of the raid force, size and type of enemy force and its mobility and state of alert, terrain and avenues of approach into the area, and the time needed to seal off the objective area. The security element may perform the following:

(a) Secure the ORP.

(b) Give early warning of enemy approach.

(c) Block avenues of approach into the objective areas.

(d) Prevent enemy escape from the objective.

(e) Provide overmatch for the units at the objective and suppressive fires for their withdrawal.

(f) Provide short-range air defense.

(3) The support element provides the heavy volume of fire needed to neutralize the objective. Because fires from this unit are violent and devastating, they must be closely controlled to ensure the precision needed. On order or as planned, fires are lifted and shifted to cover the maneuver of the assault element by suppressing enemy fire from the objective or aerial fires. The support element may also be given specific locations to cover by fire in support of the security element if an enemy quick-reaction force moves toward the objective area. These may include routes into and out of the objective site, key terrain features, or installations adjacent to the main objective. Once the assault has been completed, or on order from the raid force commander, the support element displaces to the next planned position. Organization of the support element is determined by the following:

(a) Size of the objective, the geography of surrounding area, and the enemy threat (to include air) in the area. This element should be able to neutralize the objective (when supported by air or naval gunfire) and to lift or shift fires either when the assault is launched or when ordered by the raid force commander.

(b) Mission of the assault unit.

(e) Suitable firing positions.

(d) Size and nature of the enemy force in the objective area and those enemy forces capable of reinforcement at the objective.

(e) Fire support from other units (air strikes, naval gunfire, surface-to-surface missiles, and artillery fire).

(4) The assault element seizes and secures the objective and protects demolition teams, search teams, prisoner-of-way teams, or other teams.

(a) The organization of the assault element is always tailored to the mission. Each target area must be examined carefully. To assault, seize, and destroy an installation, position, or equipment, the assault element could be organized into one or more assault teams. The element's mission is to overcome resistance and secure the objective and to destroy the installation or equipment. Other specialized teams may also be needed. For example, sniper teams could be needed to remove key sentries. To capture prisoners, liberate personnel, and seize or destroy equipment, the assault element could be organized into assault teams, prisoner teams, search teams, medical teams, demolition teams, or breach teams.

(b) To destroy a point target or installation in a heavily defended area where the USAF cannot get close en to be effective, the assault element might be organized with one small team equipped with laser target designators. From covered and concealed positions, members of the team could then provide guidance for USAF delivery of laser-guided munitions from a safe distance.

h. The site chosen for the raid LZ or DZ must support the planned actions at the objective. There are two options when choosing sites:

(1) The ranger force can land on or near the objective and seize it before the enemy can react. (See Figure 5-2.) This avoids forced marches over land carrying heavy combat loads. If there is no suitable landing area near the objective, or the enemy has a strong reaction force nearby, this option is not favored.

Figure 5-2. Airborne landing on the objective.

(2) The ranger force can land unseen far from the objective. It then assembles, reorganizes, and moves into an ORP near the objective. (See Figure 5-3.) The objective is seized after security and support elements are in place. This option makes coordinated action easier by setting up control of small units before engaging the enemy.

Figure 5-3. Airborne landing distant from the objective.

5-3. Interdiction operations.

a. AirLand Battle doctrine set forth in FM 100-5 is a shift in emphasis toward keeping the initiative in battle. The deep battle seeks out enemy elements separated from the main battle area and attacks them before they can be deployed. Interdiction of selected enemy forces by attacking them or blocking chokepoints hinders, delays, or interrupts the enemy's use of his lines of communication; it denies him the use of certain areas; it destroys industrial facilities, military installations, and equipment; and it stops or slows his advance along a particular route of approach. The ranger regiment's ability to conduct interdiction is vital to execution of the AirLand Battle doctrine.

b. Interdiction by ranger forces includes blocking or channeling approaches, holding key terrain for a limited time. and destroying facilities.

c. Ranger forces conduct interdiction missions that supplement those conducted by long-range Army and USAF weapon systems. Large-size ranger forces are normally used only when other systems fail or are not available, and the target is of such priority that it must be attacked. Small ranger laser target designator teams may work in concert with USAF or USN strike aircraft to direct terminally guided weapons systems. Rangers should be directed against crucial interdiction targets that disrupt, confuse, or delay the enemy and allow friendly forces to gain the initiative at the forward line of own troops (FLOT).

d. Typical interdiction operations are--

(1) Destroying rail lines, switches, rail yards, and rolling stock.

(2) Destroying bridges, overpasses, tunnels, and truck parks.

(3) Closing mountain passes or routes in restricted terrain.

(4) Cutting oil or natural gas pipelines and destroying pumping and distribution systems.

(5) Destroying dams, locks, spillways, and flood control gates.

(6) Destroying rail transload points, switches, and yard engines.

(7) Destroying shipyards, dry docks, wharfs, lighters, cranes, and stevedore facilities.

(8) Destroying electrical generation facilities, power lines, transformers, and grid monitoring centers.

(9) Destroying water pumping stations and filtration systems.

(10) Destroying radio and TV stations, phone lines, microwave terminals, satellite receiving stations, and telephone exchanges.

(11) Ambushing enemy forces and personnel moving along lines of communication.

e. Considerations when using rangers for interdiction are:

(1) Insertion and extraction.

(2) Support resources available.

(3) Ranger force vulnerability.

(4) Target identification.

(5) Threat environment.

(6) Other or supporting means of interdiction.

(7) Command and control.

(8) Time and magnitude of the effect on the enemy forces at the forward edge of the battle area.

(9) Collateral effects of target destruction.

f. Interdiction targets are not identified and attacked at random. They are part of an overall deep attack plan to destroy or cripple an entire system. Interdiction is based on the assigned deep campaign plan that directs which targets are to be attacked, by whom, the results desired, and the priorities of attack for specific systems. Based on his plan, the ranger force commander selects those ranger elements best suited to conduct the attack. Target selection, which requires detailed intelligence and planning, is based on:

(1) **Target criticality.** A target is critical when its destruction or damage would have a major influence on the enemy's ability to conduct or support operations. Each target is considered in relation to other elements of the particular target system designated for interdiction. The criticality of a target changes with the situation. For example, when the enemy has few locomotives, a railroad bridge is less critical.

(2) **Target accessibility.** The ranger force must be able to insert into the target area. Terrain, security controls, location, and means of insertion are considered when studying a target for accessibility. systems. Based on his plan, the ranger force commander selects those ranger elements best suited to

(3) **Target restorability.** The enemy's ability to restore a damaged facility to normal operating capacity affects target selection. The enemy's repair and maintenance ability and his reserve stocks must be considered.

(4) **Target vulnerability.** A target is vulnerable when it is open to attack by means available to ranger forces. Vulnerability is affected by the nature of the target such as type, size, disposition, and security. It is also affected by the means available to attack it such as explosives, incendiaries, and special devices.

g. Key to the interdiction process is the ability to locate and evaluate target systems for attack and the timing of the attack. A target system is a series of related functions that serve a common purpose.

(1) A target is one element, installation, or activity identified for attack such as a locomotive, train, bridge, or prison.

(2) A target complex is several targets in the same general area, such as a railway marshalling yard, an airfield, or port and dock facilities.

(3) A target system consists of an industrial system and its sources of raw materials; the rail, highway, waterway, or airway systems over which these materials are transported; the source of power and method of transmission; the factory complex; and the means by which the finished product is transported to the user. The targeting process is usually accomplished in the special operations force cell at EAC and the battle coordination element (BCE) at corps level.

h. Interdiction techniques include ambush, emplacement of conventional demolitions, target designation for laser-guided munitions, and emplacement of beacons or sensors.

NOTE: The enemy surface-to-air threat normally precludes the use of an airborne platform for direct observation and designation. This may be accomplished by the on-ground ranger force.

i. Through interdiction, ranger units are an ideal force to contribute to the Army's AirLand Battle operational concept. This includes taking the initiative, moving quickly to strike deep in the rear of the enemy, and defeating the enemy force beyond the FEBA. Planners and commanders must ensure that the ranger units are used on missions that are synchronized to contribute to the overall deep battle campaign. Strikes against critical targets must combine with efforts of other assets to take the initiative from the enemy, limiting his flexibility, initiative, and momentum.

5-4. Personnel and equipment recovery operations.

Recovery operations are specialized raids to liberate imprisoned or detained personnel or to return certain items of equipment to friendly control. These operations include recovering and extracting downed or hijacked aircrews and political or military leaders. Rangers use surprise and combat power to overwhelm resistance before harm can come to the detainees or prisoners.

a. **Planning.** The planning of personnel and equipment recovery operations is the same as for the raid. The main difference is that the ranger force commander must plan for the extraction of the recovered personnel and the loading and extraction of sensitive equipment. Personnel and equipment recovery raids are often executed as quick-response missions. They normally require the ranger force commander to plan and execute the recovery quickly.

b. **Augmentations.** This type of strike operation often requires that the ranger force be augmented with nonranger personnel with special skills or abilities. Examples are medical personnel, technical experts, mechanics or crew members trained in repair and retrieval operations, linguists or translators, and other special operations force elements.

c. **Organization.** The organization of the recovery force is the same as for a raid. Sometimes special teams are needed to perform certain missions involved in the recovery. This mission is conducted aggressively, usually ending in close combat during darkness or other limited visibility conditions. Reconnaissance teams may be inserted ahead of the ranger force to reconnoiter the objective and to locate the detained personnel.

(1) All planning and execution takes place as described for the raid. The emphasis must be on detailed, timely intelligence.

(2) Medical teams must be available to care for the detainees.

(3) The ranger personnel must be adept at searching and clearing of buildings.

(4) The ranger force must--

(a) Locate the specific building or area where the friendly detainees are being held.

(b) Gain entry into buildings, use stun grenades, and avoid obstacles and booby traps.

(c) Engage all enemy personnel using selective-kill techniques.

(d) Locate, identify, and secure detained personnel.

(e) Move personnel to a secured extraction site.

(f) Leave no detainees or rangers behind in the objective area.

d. **Execution.** The key to success is surprise and speed. The enemy must be assaulted when least expected. The ranger force must ensure that friendly detainees are not harmed during the assault - either by rangers or enemy personnel.

e. **Extraction.** The ranger force commander should plan for the extraction point to be as close as possible to the point of recovery. This prevents the detainees from having to move cross-country or for the ranger force to transport bulky equipment long distances. For small groups of detainees or small items of equipment, the extraction point can be further away. Recovery operations may use any method or combination of methods of insertion and extraction (see Chapter 4). Close planning and coordination is normally required with US Army, USAF, or USN aviation for both delivery and evacuation of the target area.

CHAPTER 6

Special Light Infantry Operations

Ranger units are normally employed on strike operations. However, broad US strategic needs demand a wide range of other military abilities. The ranger regiment may conduct special light infantry operations when larger conventional forces could not be used. The regiment conducts these operations using light infantry doctrine, methods, and techniques.

These operations include, but are not limited to, movement on foot, military operations on urbanized terrain (MOUT), passage of lines, ambush, and reconnaissance. Ranger units do not normally conduct planned operations against enemy armored forces. Battlefield surprises or METT-T considerations may cause ranger units to engage armored forces for short periods. The ranger regiment can conduct a conventional light infantry defense for short periods. A defense is not normally conducted adjacent to other friendly forces. It is normally a perimeter defense from which the unit then breaks out from encirclement and links up with friendly forces.

6-1. Movement.

a. Most missions involve movement on foot in tactical operations. Ranger units infiltrating behind enemy lines seek to avoid chance enemy contact. They choose the movement that allows them to retain security and control. To avoid loss of surprise and initiative, casualties, and mission failure, ranger units normally--

(1) Avoid chance enemy contact, if possible.

(2) Move on covered and concealed routes.

(3) Avoid likely ambush sites and other danger areas.

(4) Practice camouflage, noise, and light discipline.

(5) Maintain all-round security.

(6) Make unavoidable enemy contact with the smallest element possible.

(7) Retain the initiative to attack at the time and place of the ranger unit's choice.

b. In planning the movement of his unit, the ranger commander must consider all the elements of METT-T and determine how they will affect the movement. He must determine the route, formations, security measures, methods of command and control, location of key leaders and weapons, and action to be taken on chance enemy contact.

(1) **Routes.** The ranger force commander conducts a careful route analysis, using the factors of observation, concealment, obstacles, key terrain, and avenues of approach (OCOKA). He chooses a route that maintains the security of the force while ensuring surprise. He tries to choose a route that will avoid enemy contact and speed movement. The commander makes a detailed route reconnaissance using maps, aerial photographs, and reports from persons familiar with the area. If possible, he reconnoiters the route himself or uses a reconnaissance team. He may have that team act as guides or mark the route using covert marking devices.

(2) **Formations.** Ranger units use the movement techniques found in current light infantry doctrinal publications. The traveling, traveling overmatch, or bounding overmatch technique is chosen based on the likelihood of enemy contact. The ranger force normally tries to avoid detection and contact with the enemy while moving. The commander chooses a movement technique that ensures that if unexpected contact were made, it would be with the smallest element possible and not the entire formation. The distance between moving ranger elements depends on the terrain, visibility, and enemy situation. March interval is normally great enough to allow each succeeding element either to deploy or change direction if enemy contact is made. Distances are close enough that each element could quickly assist the element to its front.

(3) **Security.** Security is to the front, rear, and flanks when enemy contact is likely. A front security element is placed well forward with adequate communications with the main body. The security element is used to detect the enemy and warn the main body. The main body then either changes direction or hides until the enemy is past. If the enemy does detect the main body, the security element may be used to attack the enemy from the flanks or rear. Flank security elements are placed next to the route of march. They move forward either by alternate or successive bounds. An alternative is to have the security element moving adjacent to the column along routes paralleling the direction of march. Rear security is handled the same as front security. Plans are made for the rear guard to support the main body during chance enemy contact. This is done either by maneuver or by furnishing supporting fire.

(4) **Command, control, and communications.** All available communication is used, consistent with OPSEC and movement security, to assist in maintaining control during movement. March objectives, checkpoints, and phase lines may be used to aid in control. The number of reports is reduced, normally only exception reports are needed. The ranger force commander should be well forward in the formation but may move throughout as the situation demands. Communications with security elements are mandatory. Operations security often precludes the use of radios. Therefore, connecting files, runners, and visual signals are normally used. Detailed planning, briefing, rehearsals, and control are valuable if there is enemy contact. Alternate plans are made to cover all possible situations.

(5) **Weapons.** Rangers are placed within the formation to allow a mix of their individual weapons to cover the front, flanks, and rear of the formation. Rifles, grenade launchers, and light automatic weapons cover all sectors. Crew-served weapons such as machine guns, antitank weapons, and mortars should be given designated locations within the formation. This allows the element leader to employ them quickly. They should be placed near the lead element, but not so close as to be engaged along with it. Crew-served weapons are used to

provide overmatch, especially when employing the bounding overmatch technique of movement.

(6) **Rehearsals.** The ranger force, expecting movement as part of an operation, conducts the following combat drill rehearsals:

(a) Conduct movement.

(b) React to enemy contact.

(c) Maneuver under fire (fire and movement).

(d) Cross danger areas.

(e) Break contact.

(f) Conduct a hasty ambush.

(g) Make a temporary halt.

(h) Establish a perimeter.

(i) Establish a rally point.

(j) Take actions upon consolidation.

(k) Use arm-and-hand signals.

(l) React to flares and indirect fire.

c. After a ranger force is inside enemy territory, it must be alert to avoid detection before reaching the objective.

(1) Should the ranger force security element become aware of enemy presence, it alerts the main force. The main force tries to move away without alerting the enemy and continues the mission.

(2) If the enemy does make contact, the security element then engages the enemy, trying to deceive him as to the size of the ranger force and its direction of movement. After breaking contact, the security force divides into small groups and moves to a rally point to rejoin the main force.

(3) The ranger force does not fight en route to the objective unless there is no alternative, then it quickly breaks contact and moves away.

(4) Ranger units may need to set up a temporary defensive position for resupply, evacuation of wounded, planning, or extraction. One technique that can be used is the temporary perimeter defense.

(5) If the moving ranger force comes under indirect fire, it moves quickly out of the impact area and continues the mission.

(6) The most effective means of combating an ambush is through the immediate return of a

high volume of fire by all weapons. This requires alertness, discipline, dynamic leadership, and rehearsed plans.

(a) Regardless of the method of movement, all weapons should be positioned for immediate use. A well-executed ambush is hard to counteract. However, the effectiveness of an ambush can be reduced by the immediate return of fire. Enemy weapons may be silenced, thus creating weak points in the ambush position. Smoke and high-explosive (HE) grenades are effective in executing counterambush actions. They create confusion, provide a screen for movement, and disrupt the ambush plan.

(b) When all elements of a ranger force are not trapped within the ambush, the elements that are free to maneuver should initiate an immediate flank or rear assault against the ambush force. A flank assault permits better coordination with personnel in the ambush. Supporting fires are provided by those weapons that cannot be hand carried and fired in the assault.

(c) Specific actions against a near and far ambush are as follows:

- **Near ambush.** Elements caught in the kill zone return fire at once and move out of the kill zone. Elements not in the kill zone help by maneuvering and conducting a flanking attack. The unit assaults the ambush position.
- **Far ambush.** Those ranger elements not in the kill zone set up an overmatch position and lay down a base of fire for the elements in contact. The elements in contact move out of the kill zone and prepare to assault the enemy, using individual movement techniques and fire and movement.

(7) The reorganization after enemy contact involves the use of rally points, plans for local security, reorganization of the force, evacuation of dead and wounded, and movement based on the unit mission.

6-2. Ambush operations.

a. **Definition.** An ambush is a surprise attack, from a concealed position. It is used on a moving or temporarily halted enemy to destroy or capture him and his equipment. An ambush is a useful tactic because--

(1) Small, well-trained, disciplined forces, with limited weapons and equipment, can destroy much larger enemy forces.

(2) It reduces the enemy's overall combat effectiveness by destruction and harassment of his forces.

(3) Enemy morale and effectiveness suffer heavily at little cost to the force executing the ambush.

b. **Execution.** A successful ambush must be executed with precision, violence, speed, and audacity of execution. For success, ambush operations must emphasize the following:

(1) **Surprise.** Surprise, more than any other single aspect, enhances the value of an ambush. Surprise increases the potential for inflicting damage on the enemy with less danger to the

ranger force.

(2) **Coordinated firepower and shock effect.**

(a) Coordinated firepower is used for maximum shock effect. Massive volumes of accurate fire, explosives, and mines, coupled with an aggressive attack, breaks the enemy's spirit to fight back. Surprise increases shock effect and the chances for success. Shock effect can cover unexpected defects in an ambush--for example, ambushing a much larger force than expected.

(b) All weapons must be sited with interlocking fires in the kill zone and along likely avenues of entrance or exit. Mortars should be used if the terrain permits. Tripods and traversing and elevating mechanisms are normally used with machine guns to lock in fires. All riflemen use firing stakes to mark left and right limits, and elevation stakes. There is a tendency to shoot high in an ambush--especially at night. The M203 grenade launchers are sited to cover the dead space and routes of escape. When the rules of engagement permit, plastic bags containing CS crystals may be taped to the front side of Claymore mines.

(3) **Control.** Control is essential.

(a) Leaders must have contact with all members of their unit to alert them to the oncoming enemy. Leaders should not move around the ambush site during this crucial period. A method used to alert members can be to tie strings or vines to rangers' legs or arms. By a series of light tugs, all members of the ambush can be alerted to enemy presence.

(b) The leader must initiate the ambush with a casualty-producing device. A bank of Claymore mines on a double-ring main is an excellent device to spring an ambush. Other good techniques are to use a 90-mm recoilless rifle firing antipersonnel (APERS) rounds or a machine gun. All must be under the direct control of the ranger leader. Whistles or pyrotechnics must not be used. They will allow the enemy time to react.

- As soon as the enemy is hit, he reacts. The ambush force has only a few seconds to destroy the enemy in the kill zone. He will quickly recover from the initial shock and leave the kill zone--either with a direct counterattack or withdrawal, Subsequent fires and other banks of Claymore mines must be planned.
- The leader always executes the ambush, except when a member of the ambush knows he has been discovered. He then has the authority to execute--with killing fire, not by yelling.
- The cease-fire must be controlled by the ranger leader. A whistle or other device may be used to get attention, and then cease-fire is signaled.

(4) **Security.** The flanks and rear of an ambush site are open to counterattack. Flank and rear security may be enhanced by -

(a) Echeloning in depth.

(b) Designating sectors of observation.

(c) Positioning of RSTA devices.

(d) Enforcing noise and light discipline.

(e) Having a good withdrawal plan.

(f) Securing routes of withdrawal.

(g) Executing with speed and violence.

(h) Positioning security forces to sea] off the ambush area.

(i) Having good camouflage.

(5) **Simplicity.** A simple, direct plan improves the chance of success. The ambush plan must be clear yet concise to offer the greatest likelihood of success. For example:

(a) Mission statements for security, support, and assault elements must be clear, concise, and direct.

(b) Tasks to be performed by the ambush elements should be easy to understand.

(c) Contingency plans should be simple.

(d) Routes into positions and withdrawal routes should not cross. They should be the shortest, most secure routes.

(6) **Training and self-discipline.** Precise and violent execution is how ranger units win against enemy forces. All advantages must be exploited. Discipline must be strict. There must be no sleeping, talking, eating, or smoking in the ambush site. If an ambush is to be set up for long periods, then the elements of the ambush must be pulled back at set times to the ORP for rest. Extended ambushes of 24, 36, or 48 hours require six- or eight-hour shifts. It may take a company to man an extended platoon ambush position. Tired troops cannot man an ambush well. Other vigorous operations cannot be performed all day and have an alert ambush all night.

(7) **Maximum use of specialized equipment.** This includes the most sophisticated weapons, communications equipment, and RSTA devices. The effectiveness of any mission is enhanced by the use of this equipment.

(8) **METT-T.** Every ambush mission must be evaluated by the ranger commander IAW METT-T to design an ambush that works.

(a) **Mission.** Clearly state the mission--for example, stop enemy resupply, ambush enemy armor columns, or capture personnel or equipment.

(b) **Enemy.** State the type of enemy units--for example, guerrillas, main force, logistics carrying parties, or armor units. What is the size of the force, its habits when attacked, and type of support? What is the enemy night vision ability and expected time of movement?

(c) **Terrain.** Is it wooded, mountain, jungle, or urban? Some good ambushes can be set in a city or suburb. Are the routes available to the enemy suitable to his activity?

(d) **Troops.** Squads and platoons are the basic ranger ambush forces. An ambushing force should be able to engage a force two to three times its size--but evaluate enemy and terrain, A platoon ambushing an enemy tank company in daylight in open terrain is risky. But engaging that same column at night or in a forest may give the ranger platoon an advantage.

(e) **Time.** How long will it take to emplace the ambush? A complex, lethal ambush with plenty of "dirty tricks" takes hours to set up; a hasty ambush on a trail takes 15 to 30 minutes. How long is the ambush to be in place? What hours of the day, or at nighttime only? How soon must it be set up? Do not set up an ambush in darkness--even if it is to be executed in darkness. Try to allow every ranger to '6see" and "lay" in his weapon in daylight. Time is important.

c. **Organization of the ambush force.** A ranger unit conducting an ambush must be task-organized to perform the following functions: assault, security, and support. The ambush forces should be task-organized according to the TOE--by platoons, squads, and fire teams. The TOE should not be changed to create smaller elements for an ambush. The TOE formations may be reinforced with machine gun, recoilless rifle teams, or a 60-mm mortar squad.

(1) **Assault.** The elements assigned the assault mission either move directly into their positions or through a release point. The mission may include any combination of the following:

(a) Conduct the main assault.

(b) Halt an enemy's motorized column or any moving target,

(e) Kill or capture personnel.

(d) Recover supplies and equipment.

(e) Destroy vehicles and supplies.

Search teams are not always used. The ranger leader must decide how and when to use search teams. When rangers leave the security of their well-chosen, concealed ambush position, they are subject to the fires of the enemy who may also be hidden and ready. If it is at night, do not throw tripflares or shoot illumination to light the search area as this will illuminate and expose rangers to the enemy. Always assume there is hidden enemy--the ambush will not kill them all. Night observation devices (NODs) or a red filtered flashlight taped to M16s should be used to quickly search in ranger buddy teams. If the return fire from the enemy is great or if the ambush missed the main body, then the leader may choose to break contact and leave without searching the kill zone.

(2) **Security.** The elements assigned the mission of security may move to their positions

directly or by way of a release point. Their missions may include any or all of the following:

(a) Secure flanks, rear, or ORP.

(b) Provide early warning.

(c) Seal off the kill zone to prevent the enemy from escaping or reinforcing.

(d) Assist in executing the ambush.

(e) Cover withdrawal of main ambush force.

(3) **Support.** The units assigned a support mission provide fires that may include employment of--

- Heavy automatic weapon fires.
- Antitank fires.
- Mortar fires.
- Mines.
- Flame munitions.

d. **Ambush site.**

(1) **Choose an ambush site.** When choosing an ambush site, all sources of information must be used to enhance surprise, exploit the enemy's weak points, and take advantage of the terrain. An analysis of the terrain is also based on the other factors of METT-T to include maps, aerial photos, and, if possible, personal reconnaissance. Emphasis is on--

(a) Natural cover and concealment for the ambush force.

(b) Routes of entry and withdrawal (at least two) that should be direct and easy to reach.

(c) Good observation and fields of fire.

(d) Harmless-looking terrain.

(e) Few enemy escape routes.

(f) Limited enemy reinforcement ability.

(g) Nearby assembly or rendezvous area.

(h) Terrain that will canalize enemy into kill zones, and natural obstacles to keep him there.

NOTE: Try to select a site covered by friendly supporting indirect fires.

(2) **Take advantage of the terrain.** Emphasis must be on exploiting all natural cover and concealment afforded by the terrain. Site the ambush and individual positions based on the terrain rather than trying to adapt the terrain to a fixed geometric design.

(3) **Restrict enemy movement.** Restricting enemy movement by natural or man-made

obstacles should also be planned.

e. **Types of ambushes.**

(1) Ambushes have two categories for ranger purposes--area ambush and point ambush.

(a) Area ambushes may be set up by ranger platoons, companies, or battalions. They are used to interdict enemy movement in a given area or inflict casualties on his forces. Area ambushes consist of a series of point ambushes. The size and location of the ambushes are dictated by the METT-T analysis.

(b) Point ambushes are set at the most ideal location to inflict damage on the enemy. Such ambushes must be able to accept the enemy force from more than one direction. The ranger force levels may enable it to execute an ambush in two or three main directions. The other directions must be covered by security that gives early warning of enemy attack.

(2) The basic ambush is linear (Figure 6-1). An ambush is set on an expected avenue of approach. This ambush can accept contact from three basic directions--left, right, and front. The rear is secured by a security team.

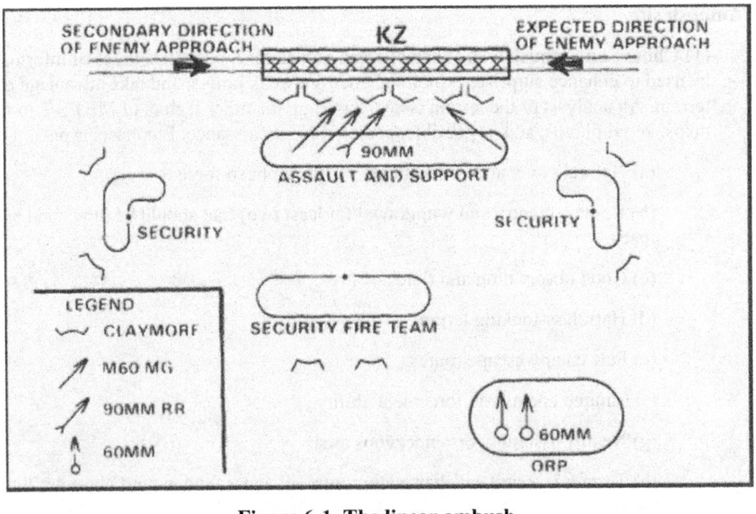

Figure 6-1. The linear ambush.

(3) A variation of the linear ambush is the L-shaped ambush (Figure 6-2). The L-shaped ambush is formed with the base (bottom) of the L perpendicular to the expected enemy direction of advance. This is a good ambush for a road, jungle trail, or an area where the enemy is canalized and his approach route is known. The L-shaped ambush can handle an enemy coming from the expected direction (toward the base of the L) from the front (stem

of the L). It is less effective against an enemy formation that comes from the opposite direction. The ambush must be executed when the enemy main body has cleared the base of the L. The flank security must protect the rear of the base of the L. In the L-shaped ambush, weapons must be sited to avoid direct or ricochet fire into friendly forces. The security elements must be considered as fire fans are being planned.

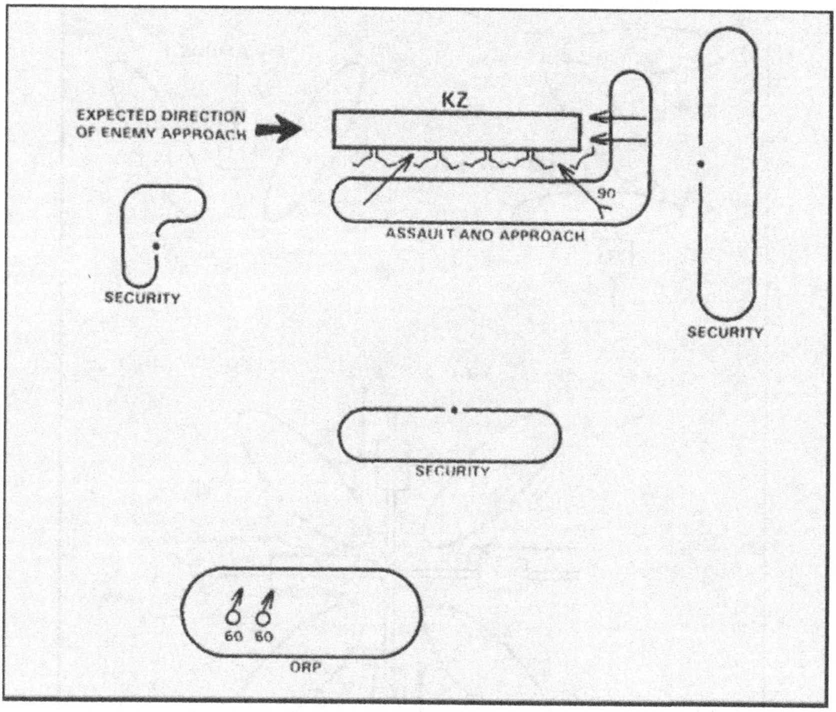

Figure 6-2. The L-shaped ambush.

(4) Other ambushes (the Z, the X, the V, and so on) are all variations of the linear and L formations (Figure 6-3). The more advanced ambush formations are difficult. Units must be expert in the basics before they are ready to employ such techniques.

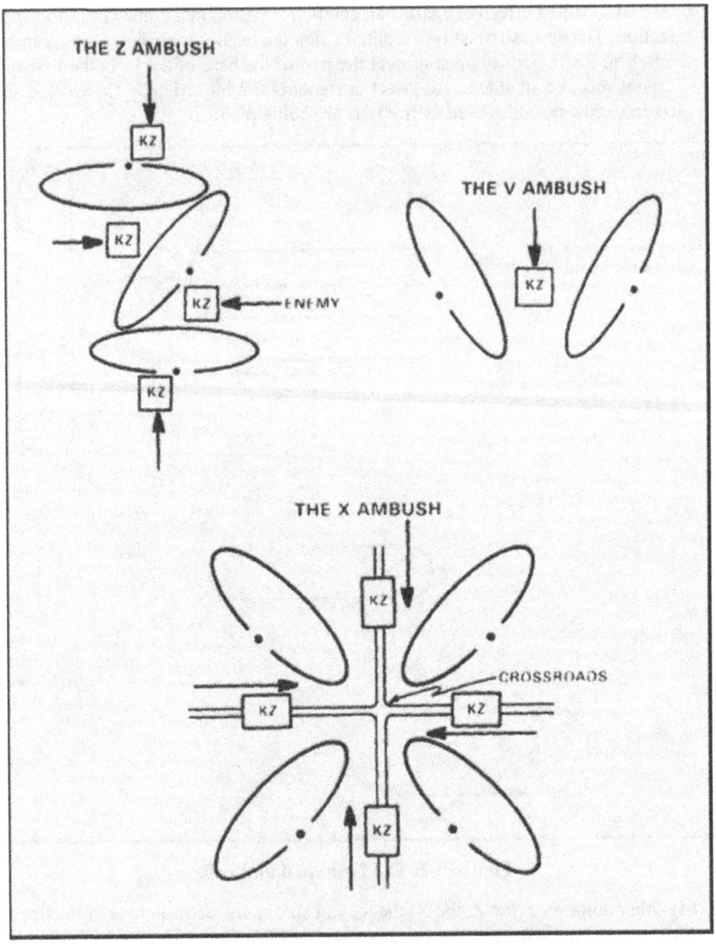

Figure 6-3. Other ambush variations.

(5) Another type of point ambush is the mechanical ambush (Figure 6-4). It consists of Claymore mines set in series with a double-ring main. It is command detonated or detonated by a triggering device activated by the enemy. Mechanical ambushes are normally manned. Rangers prepare to engage the enemy with direct fire if the mechanical ambush does not detonate or if it is wholly or partially ineffective. Mechanical ambushes are an effective way to interdict a large area using a small ranger force. If the mechanical ambush is effective and

rangers do not reveal their presence, the enemy is confused. This has a devastating effect on his morale and effectiveness.

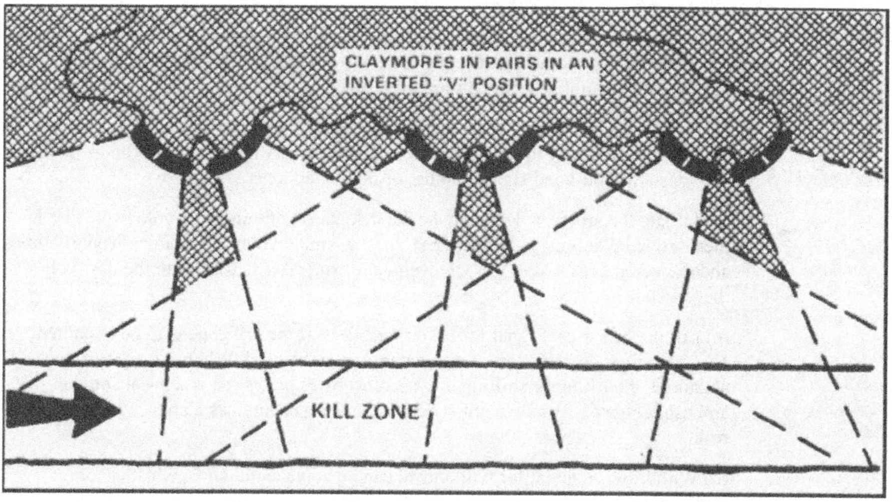

Figure 6-4. The mechanical ambush.

f. **Execution of the ambush.**

(1) **Occupation.** Stealth and security are important factors.

(a) Position security teams and early warning detection devices first.

(b) Use the rest route to main ambush position consistent with security.

(c) Quickly occupy the ambush position and set up communications and signaling devices.

(d) Position key weapons (automatic and antiarmor).

(e) Rig Claymore mines, tripflares, and booby traps.

(f) Ensure that all weapons are correctly positioned. Assign sectors of fire to provide mutual support and cover dead space.

(2) **Camouflage.** Each man must be hidden from the enemy, During preparation for the mission, each man camouflages himself and his equipment, and secures his equipment to prevent noise. At the ambush site, positions are prepared with minimum change in the natural appearance of the site. All debris resulting from preparation of positions is concealed to prevent evidence of occupation.

(3) **Movement.** Movement is kept to a minimum. The number of men moving at a time is closely controlled. Every man is as quiet as possible, especially at night. Light discipline is rigidly enforced at night. Smoking is forbidden.

(4) **Signals.** Audible and visual signals, such as whistles or pyrotechnics, must be changed often to avoid setting patterns and alerting the enemy. Three or four simple signals are needed to execute the ambush. Signals are used to--

(a) Provide early warning of an enemy approach. A signal by the security force to alert the patrol leader to the correct direction of enemy approach may be given. This includes arm-and-band signals, radio, or field telephone.

(b) Initiate the ambush. This may be the detonation of mines or explosives, Fire is then delivered at once in the heaviest, most accurate volume possible. Properly timed and delivered fires add to the achievement of surprise, as well as to the destruction of the target.

(c) Lift or shift fires, Signal for lift or shift fires, if the kill zone is to be assaulted. Voice commands, whistles, or pyrotechnics may be used. When the kill zone is assaulted, the lifting or shifting of fires must be as precise as when starting the ambush. Otherwise, the assault is delayed and the enemy has a chance to recover and react.

(d) Withdraw. A signal for withdrawal can be voice commands, whistles, or pyrotechnics.

(5) **Objective rally point.** The ORP is located far enough from the ambush site so that it will not be overrun if the enemy manages to attack the ambush unit(s). Existence loads are normally left in the ORP. Routes of withdrawal to the ORP are scouted (when possible by each man), Withdrawal routes should provide cover and concealment for the unit and hinder enemy pursuit. Withdrawal routes are a main consideration in the selection of the ambush site. They may be the key to survival after execution of the ambush. On signal, the ambush force quickly (but quietly) withdraws to the ORP. If pursued, the withdrawal may be by bounds, with mines or hasty ambushes used to delay pursuing forces.

(6) **Ambush variety.** If one method of ambush is used predominantly, the enemy will develop an effective defense against it and will be affected less by the shock of the ambush since he knows what to expect. More than one method of ambush should be used. No single method will fit all combinations of terrain, equipment, weather, and enemy capabilities. Variety should also be used when signaling - for example, use multiple signals, both audible and visual. Use weapons fire, mines, and RSTA when possible and vary signals to avoid compromise.

(7) **Swift action.** Speed in the execution of the ambush and the withdrawal should prevent enemy reaction forces from engaging the ambush force. Speed is often a shield against casualties and failure. When there is contact with reaction forces, speed may enhance quick disengagement. Ranger units must not only perform better than the enemy but also move faster than the enemy can react.

g. **Successful ambush.** Emphasize the following:

(1) Intelligence to ensure the enemy is ambushed at a time and place when he least expects or is least prepared to fight. This is preferred during reduced visibility to achieve surprise.

(2) Detailed planning and thorough training and rehearsing of all elements in all phases of the ambush. This ensures maximum shock effect through swift, precise execution.

(3) All available RSTA devices to permit daytime effectiveness at night when moving, shooting, or detecting enemy movement. To avoid detection, active RSTA devices should not be used until after the ambush has been triggered.

(4) All available firepower with emphasis on antiarmor, area and automatic weapons, and precision-guided munitions.

(5) Speed to achieve surprise and enhance security of the force.

(6) Cover, concealment, and overall protection afforded by the terrain when moving or when occupying ambush positions.

6-3. Antiarmor operations.

a. Antiarmor operations are not primary missions for ranger forces. They lack antiarmor firepower and have limited antiarmor ability. Ranger units can, however, perform limited antiarmor missions on close terrain by interdiction or ambush. The antiarmor weapons they do have are man-portable, but the ammunition is too heavy to carry many rounds cross-country. A ranger battalion has ten Dragon missile launchers and nine 90-mm recoilless rifles. It is also equipped with light antitank weapons (LAWs) and laser target designators (LTDs) that can guide Copperhead rounds.

b. Every ranger must be skilled with the limited antiarmor systems available. Proper selection, fortification, and camouflage of antiarmor ambush sites, route selections, withdrawal procedures, and initiative are important. Armor kill zones (AKZs) are normally set up after infiltrating the ranger force behind enemy lines under limited visibility.

c. A ranger antiarmor force is organized into four elements: command and control, support, security, and armor-killer teams (Dragons, LAWs, 90-mm recoilless rifles). These elements function the same as a raid force. The exception is that the assault force is the armor-killer team with the mission of destroying enemy armor, and it will probably not enter the AKZ. After the ambush occurs, the ranger force withdraws to its ORP. Indirect fires (if available) are then brought to bear on the AKZ.

d. Ranger units may conduct antiarmor ambushes and interdiction operation by use of laser-guided Copperhead rounds or air-delivered precision-guided munitions. Air-delivered family of scatterable mines (FASCAM) can be used to provide the ranger force security against attack by armored forces.

6-4. Military operations on urbanized terrain.

a. The ranger regiment normally tries to avoid the manpower-intensive, high-casualty battles associated with MOUT. However, due to the nature of strike operations and the types of targets selected, the regiment may conduct MOUT for limited periods. As in all ranger operations, the ranger force commander tries to surprise the enemy, attack him from unexpected directions, and destroy him with a violent, precise attack. The maze of cellars, rooms, sewer systems, rooftops, and alleys common to urban areas presents many opportunities for the ranger force.

b. Offensive operations by a ranger force exploit the centralized planning and decentralized execution aspects of ranger unit training. Individual action and initiative are always channeled to accomplish the commander's intent. The ranger unit's offensive tactics are based on quick action and boldness. Attacks by the ranger force are short and sharp. Execution is marked by swiftness, daring, timing, and surprise. The enemy's sleeping habits, eating habits, and relief times are watched. launched almost always at night or under smoke screens, attacks are often made without mortars or artillery, relying on surprise for success. Snipers and sharpshooters are used to shift the balance of battle.

c. Deliberate defensive operations are not good ranger unit missions. However, if a ranger unit must defend for a short period, the close terrain of urban areas allows it to function well against a stronger force. The buildings in a city break up advancing enemy formations and force them to move along streets. The ranger force executes an active defense anchored on strongpoints, such as a building or a group of buildings. It launches surprise counterattacks against the enemy's flank and rear. The aim of ranger defense is to give the enemy no respite and to make him feel threatened by a sudden, deadly attack. Ranger units should be withdrawn from defensive positions in urbanized areas as soon as possible. Prolonged MOUT and the associated casualties quickly degrade a ranger unit's combat effectiveness.

6-5. Perimeter defense.

The Perimeter defense may be conducted to control key terrain, gain time or economize forces.

a. **Planning.**

(1) **Warning order.** The order must be issued soon after mission receipt. It should contain the mission, time schedule (including rehearsal and inspection times, mission time, and place for OPORD), uniform and equipment, and special instructions.

(2) **Operation order.** The leader normally gives the OPORD on the ground to be defended. Locations and sectors of fire for squads and crew-served weapons should be designated. If time permits, leaders walk the terrain with subordinates as the OPORD is issued. The OPORD covers security, communications, signals, target reference points (TRPs), armor kill zones, and priority of engagement rules for antiarmor weapons. The OPORD also states the priority of work.

(a) Set up security and communications.

(b) Position crew-served weapons and personnel.

(c) Clear fields of fire.

(d) Prepare fighting positions.

b. **Organizing.**

(1) Squads are positioned with overlapping and interlocking sectors of fire and observation.

(2) The M60 machine guns are positioned either with an assigned final protective line (FPL) of fire or with a primary direction of fire (PDF). The PI)Fs are employed on covered avenues of approach such as gullies. The M60 FPLs should interlock across the unit front.

(3) The 90-mm recoilless rifles and Dragons are positioned with frontal cover to engage attacking armored vehicles from the flank.

(4) Riflemen are positioned to support and protect crew-served weapons. Riflemen are assigned FPLs that interlock at a point just beyond hand-grenade range.

(5) Dead space is covered with Claymore mines, antipersonnel mines, and M203 fires.

(6) Antitank mines, positioned to provide close-in protection from armored vehicles, are covered by LAW fires.

(7) Leaders plan indirect fire for long-range suppression of armored vehicles and short-range protection from dismounted assault. Mortar or artillery final protective fires (FPF) are placed on the most dangerous dismounted avenue of approach.

(8) Range cards are prepared for all crew-served weapons. Squad and platoon leaders prepare sector sketches and fire plans.

(9) The ranger force camouflages everything from ground and aerial observation.

(10) Leaders designate alternate and supplementary positions.

c. **Executing.** Leaders control the distribution and concentration of fires in their sectors. They constantly consider the overlap of fires with units to their left and right.

(1) Squads and crew-served weapons engage targets in sectors assigned by platoon leaders.

(2) Each weapon is used in its best role:

(a) Rifle, against enemy infantry within 300 meters.

(b) Machine gun, against enemy infantry within 1,000 meters.

(c) Grenade launcher, against infantry within 350 meters and armored vehicles within 200 meters.

(d) Light antitank weapon, against armored vehicles within 200 meters.

(e) Recoilless rifle, against armored vehicles within 400 meters.

(f) Dragon, against armored vehicles within 1,000 meters.

(3) Most dangerous targets are engaged first.

(4) Squads shift fires as needed or as directed by platoon leaders.

(5) The platoon masses fire on selected targets while maintaining coverage of the entire platoon sector.

(6) Platoon leaders control fires.

(7) After repelling an enemy assault, the platoon consolidates and reorganizes the defense.

6-6. Breakout from encirclement.

A breakout from encirclement is conducted when units, operating behind enemy lines, find themselves cut off from friendly forces and surrounded by superior forces. Given ranger unit missions, the chance of operating as an isolated force behind enemy lines is great. The breakout is characterized by determining enemy weak points, deception, massing of combat power, and a direct attack for a violent and timely breakout.

a. **Characteristics and fundamentals.** The breakout from encirclement involves the following characteristics and fundamentals:

(1) Set up a hasty defense. Adjust the perimeter as needed and make contact between subunits. Assign each a specific sector.

(2) Plan fire support. Plan FPFs on likely dismounted avenues of approach. Mass antiarmor weapons on the most dangerous mounted avenue of approach.

(3) Conduct reconnaissance. Identify enemy weak points. Locate enemy armored forces and the routes they must use. Locate enemy automatic weapons.

b. **Planning.**

(1) Organize all support personnel into maneuver elements--for example, company headquarters, mortars, and antitank. Organize the ranger force into a rupture force, reserve, support force, and a rear guard.

(2) Issue a fragmentary order. It should contain current local enemy situation and cover all pertinent areas of the five-paragraph field order.

(3) Select rally points.

c. **Execution.**

(1) Reduce the defense perimeter. This is done slowly until the unit is massed for the breakout attempt.

(2) Create a diversion. This is done either by fire from the rear guard force or by a limited attack in a direction other than that of the main breakout attack.

(3) Break through the defense with the rupture force. Hold the shoulders of the breakthrough.

(4) Move the reserve through to exploit the breakthrough.

(5) Move the support troops through with casualties. Have them collect casualties en route.

(6) Have the rear guard follow the support troops. It may have to fight a delaying action and use fire and movement to break contact with the enemy.

(7) Consolidate and reorganize following the breakout.

(8) Have the rupture force break contact with the enemy once the rear guard joins.

6-7. Linkup operations.

During operations behind enemy lines, ranger units may be required to link up with other friendly ground forces. Linkup may occur in airborne, air assault, amphibious, or shore-to-shore operations. It may occur during the breakout of an encircled force or during the convergence of separate forces. It may involve conventional or unconventional US forces, allied forces, or friendly guerrillas. Rangers may be required to link up with other rangers, partisans, or friendly ground forces that are conducting offensive operations.

a. **Planning.** Ranger commanders plan detailed linkup operations to avoid friendly forces firing on each other. Coordination and planning of communications, recognition signals, fires, command relationships and responsibilities, and control measures are essential. Coordination should be done before the start of the operation.

(1) Site. The linkup site should be easy to recognize, have cover and concealment, be located away from prominent terrain the enemy might use, be defensible, and provide multiple access and escape routes. An alternate site should be planned in case of compromise.

(2) **Communications and recognition signals.** Frequencies, call signs, codes, visual signals, and alternates for each of these should be planned before departure of friendly lines. Radios may be used to report the location of each unit to the site, and occupation and securing of the site. A system of mutual recognition must be devised to keep friendly patrols from firing on each other. This may include visual signals such as arm bands, panels, colored lights, or RSTA devices.

(3) **Fires.** The headquarters directing the linkup sets up fire control measures and priority of fires. The most commonly used control measure is the restrictive fire line (RFL), beyond which one force may not fire without coordination with another force. The RFL may be adjusted as two units draw closer together. Such an adjustment needs close coordination and should be planned before the operation. Successive phase lines between the two units can serve as on-order RFLS, provided that neither unit comes too close to the RFL before it is shifted.

(4) **Command relationships and responsibilities.** These are set up by the headquarters directing the linkup. Ideally, liaison personnel are exchanged before the operation. The key to success is liaison and guides.

b. **Execution.** Each member of the ranger force must understand the plan. Rehearsals are necessary in order to avoid confusion during the linkup. One uninformed ranger can start firing and cause enough confusion that friendly forces could be identified as enemy. Actions must be quick since units are open to enemy attack while they try to link up.

(1) **Moving units.** A linkup between moving units is hard. As the units move closer to each other, the chance of them engaging one another increases. Therefore, the linkup units must adjust their movements to each other and continuously coordinate on a designated secure radio net. If possible, one or both units should come to a short halt before linkup.

(2) **Stationary units.** Moving units must know the positions of the stationary units. They must keep stationary units advised of locations. Stationary units guide the moving units to the linkup point by radio. The stationary units must be ready to accept and position the moving units smoothly and quickly.

6-8. Passage of lines.

A passage of lines is an operation in which one unit moves through another unit that is stationary and disposed in a tactical formation on a FEBA. Ranger forces may conduct a passage of lines to get behind the enemy. Movement in forward unit areas must be controlled, coordinated, and kept to a minimum. This avoids conflict with friendly troops or the activation of RSTA devices. Ranger units treat the positions of forward units as danger areas. They must be assumed to be under enemy surveillance in all weather or visibility. The ranger force is open to attack during the passage. Personnel and units may be more concentrated than is desired. Fires of the stationary unit may be masked for a limited time. The passing ranger unit may not be in a good position to react to enemy contact. Detailed reconnaissance and coordination are crucial to ensure that the passage is conducted quickly and smoothly.

a. **Principles.**

(1) **Coordination.** The ranger force commander, or his representative, must coordinate departure and reentry of friendly front lines (FFL). This is done either directly with the units through which the rangers pass or through designated staff agencies. Items to be coordinated include fire support, passage lanes, control measures, guides, communications, intelligence, CSS, and actions on enemy contact.

(2) **Reconnaissance.** The ranger force commander should make a ground reconnaissance of the area through which the unit will pass. He should pick the initial rally point (IRP) at this time. The reconnaissance should include observing the area just forward of the friendly unit for routes, danger areas, and obstacles. During reconnaissance or coordination, rangers should wear the same uniform as the troops in the forward unit. This will avoid attracting the enemy's attention.

(3) **Security.** Maintain security to avoid contact with the enemy while departing or reentering. It is important that the ranger force make no enemy contact at this time. Control is hard if a fire fight starts just forward of an FFL.

b. **Departure of friendly front lines.**

(1) Designate an IRP.

(2) Maintain security. Use an appropriate movement formation when departing the FFL.

(3) Do not move within the friendly unit's area without a guide.

(4) Make final coordination with the friendly unit commander.

(5) Count the ranger force members as they depart.

(6) Make a security and listening halt.

c. **Departure of friendly units.**

(1) The ranger force arrives at the forward unit and is met by a guide. The guide leads the rangers to an assembly area chosen by the ranger commander during initial coordination with the forward unit commander.

(2) No one should move, either singly or as a unit, anywhere in the FFL without a guide.

(3) The ranger commander makes final coordination with the forward unit commander. He determines if any changes have taken place since initial coordination and learns of any recent enemy activity.

(4) Before leaving to coordinate, the ranger commander gives instructions (called a contingency plan) for what should be done while he is gone. These instructions state where he is going, who is going with him, how long he will be gone, what to do if he does not return, and actions to be taken if there is enemy contact.

(5) On the commander's return from final coordination, he may issue a fragmentary order (FRAGO) to cover any changes.

(6) Three common threats and techniques for countering them during departure from the FFL are:

(a) **Ambush and chance contact.**

- The ranger force commander must avoid enemy contact while departing.
- The ranger force goes to the friendly side of the departure point where a point team is dispatched to see that the area forward of the barriers is clear.
- The point element should check the area out to the first covered position or an area large enough to allow the unit room to maneuver if it is hit while departing.
- The distance varies with the size of the ranger force and the terrain.
- The point team notifies the ranger commander when the area is clear. The rest of the unit is brought forward.

(b) **Indirect fire.** If enemy indirect fire is falling, the ranger commander should not halt his unit after final coordination at the forward command post. Instead, he should move it quickly through the FFL. The point element should be far enough ahead of the main body to provide security from ambush or chance contact.

(c) **RSTA.** If the enemy has radar, sensors, or night vision devices, they may be offset by the following countermeasures:

- Use a well-hidden departure point, such as a reverse slope or a dense woods.

- Infiltrate the unit through the departure area and have the rangers rendezvous in a defilade.
- Pass through during rain, fog, snow, or darkness to help conceal the passage.
- Employ electronic countermeasures.

(7) Have a security and listening halt after the ranger force has moved out of sight and sound of the FFL. This is a short halt to see if enemy is in the area, and to accustom the unit to the sights and sounds of the battlefield. Halt in a position that covers the unit from chance friendly small-arms fire. If the security and listening halt must be in an exposed position, have each man lie flat on the ground.

(8) During movement, have frequent halts to see that the unit is not being followed and that no enemy is in the area.

d. **Reentry of friendly front lines.**

(1) Set up and occupy a reentry rally point (RRP).

(2) Maintain security at the RRP and at the reentry point.

(3) Use as few personnel as possible to locate the reentry point while the bulk of the unit remains in the RRP.

(4) Meet a guide at the reentry point. Normally, a forward-of FFL password is used since it may be overheard by enemy.

(5) Count the ranger force members to ensure no infiltrators follow the unit, especially during reduced visibility.

(6) Give the friendly unit commander a spot report, providing combat information of immediate tactical value.

e. **Reentry of friendly units.**

(1) The unit is moved into a rally point near the reentry point. This rally point should be on a prominent terrain feature where the ranger commander can pinpoint his location, with respect to the reentry point.

(2) The forward unit is alerted by radio that the ranger unit is ready to reenter. A code word is used for security and brevity. The code word must be acknowledged by the forward unit before the ranger force reconnoiters the reentry point. This shows that a guide has been sent to the reentry point and is waiting for the unit.

(3) If the ranger commander is certain of the reentry point location, he moves the entire unit directly to the reentry point. He has the point element coordinate the reentry.

(4) An effective method of locating the reentry point is through the use of a thermal imagery device. It detects the body beat of the reentry guide, even though he may be well camouflaged or his location masked by smoke. If ground surveillance radar is available, it can be used to vector the ranger unit to the reentry point.

(5) The reconnaissance party should never move parallel to friendly barriers or move around the wire. If the reentry point cannot be found by the initial reconnaissance or if the reconnaissance party finds only the barrier wire, the unit should notify higher headquarters and move to another rally point to wait until daylight. The unit should not stay in a rally point from which a radio transmission has been made.

(6) When the reentry point is found, the ranger commander can bring the unit forward. If the reentry point is easy to find, he can call by radio for the unit to come forward.

(7) The guide leads the unit through the barriers to the security position coordinated with the forward unit commander. The unit halts in the security position, remaining in movement formation. The ranger commander then gives a spot report to the forward unit commander. The ranger commander tells him anything of intelligence value or of immediate tactical use to the forward unit.

6-9. Reconnaissance platoon operations.

a. The reconnaissance platoon organic to the ranger regiment consists of a platoon leader, a senior reconnaissance sergeant, and three five-man reconnaissance teams that include a communications specialist in each team (see Appendix A). This platoon is responsible for short-duration reconnaissance of objectives for the ranger force commander. This reconnaissance is routinely in support of future strike or special light infantry operations. The teams can also be used by the ranger force commander to conduct deep target acquisition and designation, or to confirm or disprove information reported by other sources. Rarely will teams be used to attack enemy targets.

b. The reconnaissance teams of the ranger regiment do not replace (or perform the same reconnaissance tasks associated with) long-range reconnaissance patrol (LRRP) units or long-range surveillance units (LRSUs) that may be organic to the corps or division. These units are organized, trained, and equipped for long-term passive surveillance and reconnaissance in enemy territory. The ranger reconnaissance teams are trained and organized to support the immediate intelligence-gathering needs of the ranger force commander. They are not to act as a human intelligence (HUMINT) collection asset for other operational headquarters.

c. The reconnaissance platoon headquarters trains, equips, and controls the employment of the three reconnaissance teams. It operates the communications stations needed to receive reports from deployed teams. In the absence of the S2, it debriefs team members upon extraction or exfiltration. It can provide one or more teams to support each ranger battalion on independent operations. The reconnaissance teams can--

(1) Infiltrate the objective area by parachute (HALO, HAHO, or static lines), helicopter, fixed-wing aircraft, SCUBA, small boat, foot, or other means.

(2) Remain undetected in the vicinity of the objective area up to five days.

(3) Perform reconnaissance operations employing a full range of night observation devices, infrared detection devices, unattended sensors, and photographic equipment.

(4) Perform demolition target analysis.

(5) Operate small watercraft and inflatable boats.

(6) Emplace unattended ground sensors, omni-directional navigational beacons, hand-emplaced expendable jammers, and electronic target designation devices.

(7) Collect combat information to satisfy priority information requirements and mission-essential elements of information. Teams report that information by use of long-range, secure, burst-transmission communications equipment.

(8) Perform DZ selection, marking and reception duties.

(9) Report objective area weather conditions.

(10) Perform highly selective, limited attacks or ambushes when so tasked.

(11) Link up with the main body of the ranger force in the objective area, or escape and evade the enemy in order to return to friendly lines.

(12) Act as part of the AAT during airborne operations.

d. There are many limits of the reconnaissance teams.

(1) Mobility is normally restricted to foot movement in the area of operations.

(2) Use of radio and other active electronic or optical devices makes the teams open to enemy detection.

(3) All supplies and equipment needed by the team must be carried on the initial insertion, since resupply increases the probability of detection by the enemy.

(4) Conduct of insertion and extraction operations deep in enemy-held territory is difficult due to the need for secrecy, timeliness, security, and accuracy of location.

(5) Medical treatment of team casualties is limited to individual first aid. Casualty evacuation is difficult unless linkup with the main ranger force has been achieved.

e. The training of reconnaissance team members parallels that given to all members of ranger units with emphasis on infiltration, communications, and reconnaissance skills. Extensive training is also given in survival, evasion, resistance, and escape (SERE) and advanced first aid. Training emphasis is placed on actions at the reconnaissance site. The reconnaissance teams must conduct their mission and remain unseen. Every team member is trained to consider the need for information balanced by the likelihood of detection. Detection by the enemy compromises future operations.

6-10. Reconnaissance planning.

a. Detailed planning ensures that the reconnaissance mission is successful. It is also vital to the survival of the reconnaissance team. This planning is conducted by the ranger force planning group assisted by elements of the JTF staff. The supporting air or naval commander, navigator, and electronic warfare officer may aid the ranger force staff in the planning.

b. Upon receipt of the mission, the team selected begins an intense preparatory phase at the REMAB or the ISB. There the team receives its initial mission briefing. The items covered allow the team to complete its detailed planning. The reconnaissance platoon leader reviews the team plan. He ensures compatibility and integration into the ranger force commander's plan for employment of the main force. Staff specialists at ranger regiment and battalion--such as those concerned with intelligence, weather, or communications--provide help.

c. Planning for the use of reconnaissance teams is conducted within the framework of the ranger planning sequence for the entire mission. There are five planning considerations for the employment of reconnaissance teams:

(1) Operations security must be considered in all staff efforts to provide maximum protection for the reconnaissance teams. This includes intelligence, CE, logistics, administration, and maintenance. Operations security must be integrated throughout every ranger force operation. It consists of four main categories of security measures: signal security, physical security, information security, and deception. All are related and must be considered at the same time for each operation.

(2) Some missions may require rapid deployment to the operational area, thereby dictating the quickest method of insertion. However, others may depend on maintaining total secrecy, regardless of time.

(3) The transportation means selected for the delivery and recovery of the teams depends on the mission. This selection is also based on the abilities, limits, and availability of mission support platforms.

(4) The distance to and from the objective area must be considered, as well as the distance from the insertion point to the actual target. Reconnaissance teams can move great distances over rugged terrain. However, this movement does require time and limits the team's ability to carry bulky or sensitive equipment.

(5) Consideration must also be given to intelligence information available concerning the objective area which includes, but is not limited to--

(a) **Enemy situation.** The enemy threat--his abilities, disposition, security measures, detection, and air defense systems - affects the selection of the means for delivery or recovery.

(b) **Terrain.** Land formations must be considered in selecting the method of infiltration and exfiltration. Terrain affects the selection of altitudes, approach and exit routes, landing areas, drop zones, and beach landing sites. Air infiltration routes that provide terrain masking are desirable in static line parachute operations.

(c) **Weather.** Seasonal weather conditions affect team delivery and recovery. Factors to be considered are temperature, precipitation, visibility, clouds, and wind. If para-SCUBA techniques are used, high surface winds and their effect upon surf conditions or period of reduced visibility may preclude the use of parachutes, inflatable boats, or swimming as entry or recovery techniques. These same conditions favor land infiltration and exfiltration, The AWADS reduces the impact of visibility as a limiting factor for air infiltrations.

(d) **Astronomical conditions.** Periods of sunrise and sunset, moon phase, moonrise and moonset, and twilight must be considered.

d. A reconnaissance team mission must be specific and support the mission. Information concerning the target may not be gathered because the risk of detection by the enemy is so great. Detection may preclude the conduct of the planned operation. If more than one mission is assigned, priorities are set, All team missions must be coordinated to avoid other friendly forces operating near the target. Ranger forces conduct operations at such depth behind enemy lines that no other US or allied forces are normally present. There is a chance that a friendly indigenous guerrilla force may be operating near the objective, but ranger reconnaissance teams do not normally make contact with them.

e. The reconnaissance platoon leader joins with the intelligence and operations sections of the ranger force headquarters in the initial planning for reconnaissance team operations. The method of operation while moving, communication procedures, reporting, surveillance procedures, and actions on enemy contact should be included in the platoon's SOPS.

f. The ranger force intelligence officer normally prepares the reconnaissance plan IAW the guidance provided by the controlling headquarters. He is helped by the operations officer and the reconnaissance platoon leader. Selected team members and a member of the unit providing the insertion and extraction means are briefed on the mission early in the planning phase. They should participate in the detailed planning to follow. The essential details of a reconnaissance plan normally include--

(1) Areas to be observed and recommended areas or positions from which this can be done. These positions are determined in advance of the insertions of teams. They are based on a study of terrain, road and rail nets, enemy order of battle, delivery means, operations plans of the controlling headquarters, and the desires of the ranger force commander.

(2) Load plans and procedures. This includes both insertion and extraction. Alternate sites are chosen for possible use. Primary and alternate DZs are selected, if applicable.

(3) The flight plan. This includes insertion and extraction flight routes and alternate routes.

(4) The movement plan to and from a patrol base. This includes the times, routes,. and destinations when individuals or small groups are required to operate away from the patrol base.

(5) The fire support plan. This includes the location and times of planned air strikes in support of the main body, and all planned fires by the main assault force. The reconnaissance teams should know the frequencies and call signs for tactical air support.

(6) The timing and ground tactical plan for execution of the main force operation. This includes disposition of the main force, special mission teams from the ranger force, or any other special operations force elements.

(7) The communications plan. This includes frequencies, reporting schedule, emergency reporting procedures, and alternate communications plans. It also includes the ground-to-air

communications plan and the friendly recognition signal to be used during linkup with the main assault force or other friendly forces.

(8) Plans for use of technical specialists or special equipment.

(9) The plan for the treatment and evacuation of sick or wounded team members from the objective area.

(10) The E&E plan. This includes routes out of the objective area, passwords, and alternate extraction zones or methods far from the objective area.

(11) Mission-specific elements of information desired by the ranger force commander, and any special or on-order missions the team must conduct.

g. The team leader uses specific steps in planning, preparing, and executing missions. These procedures are comprehensive, yet adaptable to any situation. Three of the most important steps are:

(1) **Issuance of warning order.** This action gives the team maximum time to prepare for the mission. It should be issued as soon as the team leader has been notified of a mission and updated after he has made his tentative plan.

(2) **Issuance of OPORD.** The team leader conveys the information and instructions needed to accomplish the mission. He requires a briefback from each team member.

(3) **Rehearsals.**

h. The specific steps used when planning missions are as follows:

Step 1. Study the mission.

Step 2. Plan use of time.

Step 3. Issue initial warning order.

Step 4. Study terrain and situation.

Step 5. Make tentative plan.

Step 6. Organize.

Step 7. Update warning order.

Step 8. Coordinate (continuous throughout).

Step 9. Consult aerial photographs and map updates.

Step 10. Complete detailed plan. (To be combined with paragraphs 3, 4, and 5 of the operation order.)

Step 11. Issue operation order.

Step 12. Supervise (throughout), inspect, rehearse.

Step 13. Execute mission.

i. A workable E&E plan must be developed. Such a plan enhances the survival of the reconnaissance team members. It ensures that they are psychologically prepared to conduct the mission, by knowing there is a system and plan for returning them to friendly lines. The E&E plan must include a section describing the actions of the reconnaissance team members should their insertion craft be disabled en route to the objective area. It should also include what to do if the team is unable to link up with the main ranger force, or if the main ranger force has not been inserted into the objective area.

j. Rehearsals and briefbacks are best for finding flaws in procedures or planning. Thorough coordination of all procedures is essential. The exact type of aircraft or other means that will be used for insertion should be used for training. Rehearsals should be conducted under terrain, astronomical, hydrographical, and meteorological conditions close to those to be met on the operation. The team leader conducts a briefback with the platoon leader. This lets the platoon leader decide if the team is ready for the mission. It also allows him to make changes.

6-11. Execution of insertion.

a. Reconnaissance teams normally conduct blind insertions. Blind insertions are amphibious, heliborne, or parachute drops onto unmarked DZs without help from a DZ-marking party or a CCT. Blind insertions often depend on favorable light and weather conditions in the objective area. Favorable conditions for the reconnaissance team often means low-lying cloud cover, fog, and darkness. The teams are able to operate during reduced visibility by the extensive use of night observation devices. The team's extensive training and high level of land navigation skills allow it to quickly traverse rugged terrain and not be detected.

b. Team insertion into the objective area often needs a cover and deception plan. Team members can conduct HALO and HAHO operations. Teams can be inserted into areas not normally acceptable as a DZ for conventionally trained parachutists. By the use of HAHO techniques, reconnaissance teams can assemble in the air and glide to a landing at a location far from the drop point.

c. If reconnaissance teams need special or bulky equipment that cannot be carried on the individual team member, it can be airdropped, hidden near the DZ, and retrieved later. Such container drops are often used during low-level static line insertions. The high-altitude airdrop resupply system (HAARS) can be used to deliver such items at the same time as reconnaissance teams are inserted by HALO.

CHAPTER 7

Combat Support

Combat support is used to enhance the combat power of the maneuver companies. Combat support elements may be organic or nonorganic. They can conduct a variety of missions in providing indirect and direct fire support, security, intelligence, and maneuver and fire control.

7-1. Intelligence support.

a. The success or failure of a ranger mission often depends on the accuracy, detail, and timeliness of intelligence information. Both deliberate and quick-response missions need specific information that can be obtained only by a multidiscipline collection and analysis agency. An active interface with all levels of the intelligence system helps get the needed information quickly.

b. The theater or JTF commander is responsible for providing intelligence support. The 1st SOCOM also provides information and analyzed intelligence to the ranger force commander. The main source of processed intelligence is the Echelons Above Corps Intelligence Center (EACIC), similar to the division level military intelligence (MI) all-source production center (ASPC). The EACIC provides intelligence, security, and electronic warfare (EW) interfaces (see Figure 7-1). It is through the EACIC that ranger commanders can request support from national, joint, combined, or CONUS intelligence analysis systems.

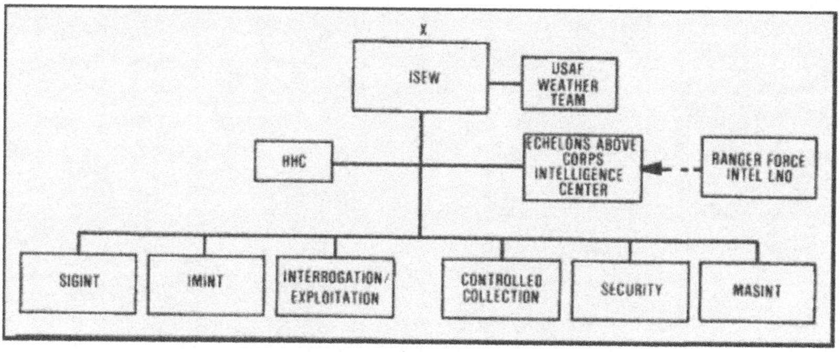

Figure 7-1. Intelligence at echelons above corps.

c. Intelligence requirements for ranger operations are unique, highly sensitive, and compartmented. The intelligence officer of the 1st SOCOM and the appropriate 1st SOCOM regional planning cell coordinate with theater intelligence agencies to provide needed information.

d. The ranger regimental staff and the ranger battalion staff have personnel in their intelligence sections to correlate and analyze information from many sources.

e. The reconnaissance platoon, organic to the ranger regiment, maintains surveillance of the objective area before the insertion of the main ranger force. It reports any last-minute developments that would affect the ground tactical plan. The platoon can also be tasked to determine any mission-specific items of essential information that the ranger force commander needs to complete his plan. This platoon is a highly specialized unit that is reserved for use by the ranger force commander. It is not a HUMINT collection agency for use by the theater or JTF commander. Reconnaissance teams normally link up with the ranger force in the objective area and are extracted with the main body. If this linkup cannot be made, they will exfiltrate enemy territory. They may use escape and evasion nets set up by other special operations forces or Department of Defense (DOD).

f. The ranger regiment normally places a liaison team at the headquarters of the command having OPCOM of a ranger force. This team provides command and control communications and liaison. It puts an intelligence liaison officer (LNO) at the EACIC or the appropriate ASPC (see Figure 7-2). This provides coordination to ensure that the intelligence needs of the ranger force are relayed to the intelligence processing center, and that the analysis is on the specific needs of the ranger force commander. This LNO coordinates with another located at the corps or EAC targeting center. Close coordination between the regiment, the intelligence analysis center, and the targeting center is essential to effective employment.

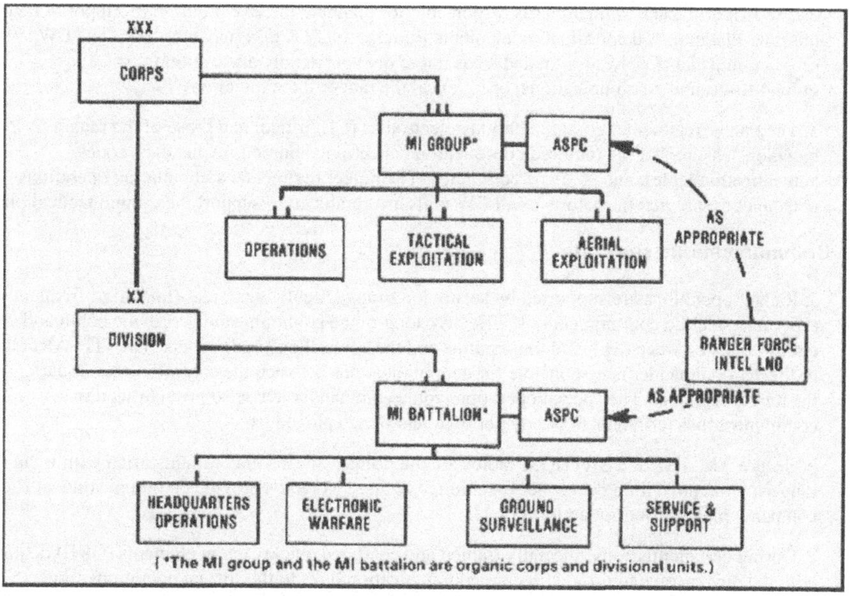

Figure 7-2. Intelligence source at corps and division.

7-2. Electronic warfare support.

a. Depending on the nature of the target and enemy abilities, a ranger mission may need support from EW units. This is mainly true during insertion. Electronic warfare operations disrupt or destroy the enemy's command and control of his forces and weapons systems, and retain friendly use of the electromagnetic spectrum. Electronic warfare is also used to support deception operations. They mislead the enemy by manipulation, distortion, or falsification of indicators, and persuade him to react in a manner prejudicial to his own interests.

b. Active jamming and chaff dispersal can prevent enemy early-warning radar from detecting the airborne force or determining its route. False transmissions can aid deception plans or feints used to help the ranger force's entry behind enemy lines. Selective jamming and imitative deception can disrupt the enemy command and control nets. This prevents the enemy from reacting to the presence of the ranger force in time to prevent mission accomplishment, or slows and interrupts his deployment of reaction forces.

c. Ranger elements, or other military forces, may conduct closely coordinated direct attacks against enemy EW sites during a ranger force's insertion or extraction. Not only does this reduce the enemy's ability to direct EW against the ranger force command, control, and communication system, but it misleads and confuses the enemy as to the true ranger objective.

d. The JTF or theater commander is responsible for planning the EW program to support a ranger mission. Planning and coordinating elements from the 1st SOCOM may also help. The EW program must be closely coordinated so as not to disrupt friendly air-to-ground or ground-to-ground communications, or to reveal the ranger force intentions.

e. The ranger regimental communications-electronics (CE) officer and those of the ranger battalions ensure that the required coordination concerning jammed frequencies, codes, authentication tables, and so on, is completed. The ranger regimental and battalion operations personnel ensure that the efforts of all EW units are combined to support the ground tactical plan.

7-3. Communications support.

a. Ranger operations are supported by secure long-range, lightweight, real-time, high-frequency (HF), and satellite communications. Effective long-range communications provide command and control links between deployed ranger units and the controlling headquarters. The JTF, ARFOR, or theater commander is responsible for communications between the controlling headquarters and the ranger regiment. The special operations force commander may also provide secure communications terminals to the ranger regiment or a deployed battalion.

b. Secure AM, FM, and SATCOM radios are the primary means of communication within the ranger regiment. Within the ranger battalion, AM and FM radios provide communications to company, platoon, and squad level.

c. During certain missions, specially trained and equipped quickreaction elements (QRE) deploy with the ranger force to provide secure communication links to the special operations force commander. These JCSE elements operate on either SATCOM or tactical satellite terminal (TACSAT) channels or through an airborne communications relay platform (see Figure 7-3). Depending on the mission, an airborne battlefield command and control center (ABCCC) aircraft or a joint airborne communications center/command post (JACC/CP) may be used. Their communications systems can operate at all levels of the national chain of command to permit a quick response to the tasking authority.

Figure 7-3. Airborne communications relay platform.

d. When only one ranger battalion is employed and the regimental headquarters is not the controlling headquarters, the ranger regiment provides a liaison cell to the controlling higher headquarters. This liaison cell includes a communications element from the regimental communications platoon. It can provide secure SATCOM, teletype, and facsimile transmission support to the ranger force in the objective area.

e. If two or more ranger battalions are employed, the ranger regimental headquarters deploys and acts as the command and control headquarters. The regimental communications platoon would then provide another communications link to the special operations force commander.

f. The ranger regimental CE officer ensures that the necessary communications links are set up and coordinated. The many communication means and channels available provide for effective control of a deployed ranger force. However, communication means must be closely coordinated at all levels of command to control the complex operations of a ranger mission. Planning and coordination with supporting aviation, transport, fire support, medical, and logistical elements before an operation are vital to efficient communications. The regimental CE officer must consider the communication systems linking the ranger force and other services. Air-to-ground and ship-to-shore communications are vital and must be set up early in ranger operations.

7-4. Fire support.

a. **Planning.**

(1) The ranger regiment has limited organic fire support assets. The six 60-mm mortars in each ranger battalion (two in each rifle company) provide the only organic indirect fire

available to the battalion commander. Each rifle company also has three 90-mm recoilless rifles and three Dragon missile systems that provide a small measure of organic direct fire support.

(2) The missions assigned to the ranger regiment and its subelements normally require fire support from organizations outside of the regiment. To plan, coordinate, and control these fires, the ranger regiment is organized with personnel dedicated to these tasks.

(3) The ranger regimental headquarters has a fire support element consisting of one fire support officer (FSO), two noncommissioned officers, and two fire support specialists. They prepare the fire support annex to the regiment's operation order. They ensure that all available fire support is planned to best support the regimental commander's ground tactical plan. They act as the primary coordinator for all external fire support, such as artillery, armed helicopter, close air support, naval gunfire, and short-range air defense. They also coordinate with each ranger battalion FSO to ensure unity of effort. The fire support element coordinates the efforts of air force, naval, and marine liaison officers when they are attached.

(4) The regimental commander may assign part of the fire support element to a committed ranger battalion as an attachment to help in the integration and planning of fires. The ranger battalions request through the fire support element that more or different fire support means be integrated into the ground tactical plan.

(5) The ranger battalion headquarters has a fire support team headquarters consisting of the battalion's FSO, two noncommissioned officers, and two fire support specialists. The FSO is the prime fire support coordinator for the ranger battalion commander. He maintains coordination with the regimental FSO and any attached fire support coordination elements, such as the USAF tactical air control party or naval shore fire control party. The FSO performs the following:

(a) Advises the ranger battalion commander on all fire support matters.

(b) Recommends allocation of fire support.

(c) Prepares fire support plans.

(d) Assigns target numbers.

(e) Processes target lists from the ranger company fire support teams to eliminate duplication.

(f) Monitors and functions as net control on the ranger battalion fire support coordination (FSCOORD) net.

(g) Reports changes in the status of fire support units to the ranger battalion commander, staff, and FISTS.

(6) When the ranger battalion is operating a single TOC, the FSO and his element are normally located there. When the ranger battalion is operating two TOCS, one TOC is normally manned by the FSO, the fire support sergeant, and a fire support specialist. The

senior fire support sergeant and the other fire support specialist would then monitor the FSCOORD net at the other TOC. The FSO ensures that any attached fire support elements also divide when needed to operate in a two-TOC configuration.

(7) The ranger battalion fire support team headquarters has three fire support teams assigned to it. These are normally allocated one to each of the ranger rifle companies. The company FSO and his team--

(a) Locate targets and request and adjust surface-to-surface fire support (mortar, field artillery, and naval gunfire).

(b) Plan fires to support the company ground tactical plan and prepare target lists.

(c) Coordinate fire support requests through the FSO for surface-to-surface and air-to-surface fires.

(d) Report battlefield information.

(e) Control air-to-surface fires of helicopters and prepare to assume control of fires from fixed-wing aircraft.

(f) Coordinate airspace use within the ranger rifle company operational area.

(g) Advise the ranger company commander on all fire support capabilities, limitations, and methods of employment.

(h) Inform all fire support units of target priority changes.

(i) Designate targets for laser-guided munitions.

(j) Assume operational control of, and administrative responsibility for, all augmentation fire support personnel (TACP, shore fire control party, and attack helicopter liaison teams).

(8) The company FSO is normally with the ranger rifle company commander. The FOs and their RATELOs are normally with the ranger rifle platoon leaders.

b. **Coordination.**

(1) Coordination of fires is a continuous process. It involves implementing the fire support plans of the ranger force commander and reacting to changing tactical situations. The success of a mission often hinges on how well fires are coordinated during the operation.

(2) The fire support planners within the many echelons of the ranger force become fire support coordinators during the actual execution of the operation. They use standard fire control coordination measures.

c. **Organic fire support.**

(1) **Indirect fires.**

(a) The 60-mortars constitute the only organic indirect fire support available within the ranger regiment. The mortar squads in each ranger rifle company normally operate as a section under the control of the mortar section sergeant. However, they can operate alone for short periods.

(b) The 60-mm mortar is used to destroy enemy equipment and light vehicles and to attack enemy personnel. Its main advantage is the ease with which it can be brought into operation against the enemy. This ease of employment allows the ranger force commander to concentrate combat power at the critical time and place. This helps to surprise and destroy the enemy.

(c) Mortar fires are employed by the ranger rifle company commander and coordinated by the company FSO. They are normally requested and adjusted by the FIST observers. The mortars are normally emplaced in a central position, and their fires are controlled by the fire direction center (FDC) using the indirect-fire method. The mortars may also use the direct-alignment or the direct-lay method.

(2) **Direct fires.**

(a) Each ranger battalion is authorized 90-mm recoilless rifles and Dragon missile launchers. These weapons constitute the heaviest organic direct-fire ability in the regiment. The antitank section in each ranger rifle company may operate as a complete section, but normally individual weapons are assigned throughout the company sector wherever their fires can best support the mission. These weapons can destroy most enemy armored vehicles and can penetrate the flank and rear armor of all enemy main battle tanks. The 90-mm recoilless rifle has a multiple projectile round that is effective against exposed enemy personnel. The HE round is highly effective against buildings and bunkers.

(b) The main disadvantage of direct-fire antiarmor weapons is their weight. These weapons, and their ammunition, are man-portable but place a strain on the unit when moving across rugged terrain.

(c) The ranger battalion is trained in the use of light antiarmor weapons, such as the US M72A2 LAW and the Warsaw Pact rocket-propelled grenade (RPG). The following provide protection for the ranger force against attack by enemy armored forces:

- The employment of light antiarmor weapons at all organizational levels.
- The placement of the 90-mm recoilless rifles or guided-missile systems on the most likely avenues of enemy-armor approach.
- An aggressive, offensively oriented employment.
- The ability to move into and across terrain impassable to armored vehicles.

d. **Field artillery fire support.**

(1) Many ranger unit operations are out of the range of supporting field artillery fires. Whenever such fires are available, they are planned for and integrated into the ranger force ground tactical plan.

(2) Coordination and exchange of call signs, frequencies, and target lists occur before insertion of the ranger force. Unless required by the tactical situation, ranger units do not normally receive augmentation forward observers from supporting field artillery units. The degree of mental and physical training necessary to execute certain ranger techniques requires more preparation than field artillery units are able to provide. The FIST from the ranger battalion requests and adjusts field artillery fires for the ranger force.

(3) Field artillery fires can be used to support the ranger force even if the objective area is out of range. Field artillery cannon fire and multiple rocket launch system (MRLS) fire can be used to suppress enemy air defenses. This helps the ranger force as it crosses the FEBA during airborne or air assault operations.

(4) Field artillery fires can be used to support the exflltration of ranger elements as they approach friendly lines. These fires can contribute to the deception plan and add combat power to feints used to support ranger operations.

(5) When in range of the objective area, field artillery units can be used to emplace field artillery delivered FASCAM to enhance the security of the ranger force.

(6) Copperhead rounds fired by 155-mm field artillery units can be terminally guided by the ranger FIST forward observers (FOs). They can attack hardened point targets or enemy armored vehicles by using a man-portable laser target designator (see Figure 7-4).

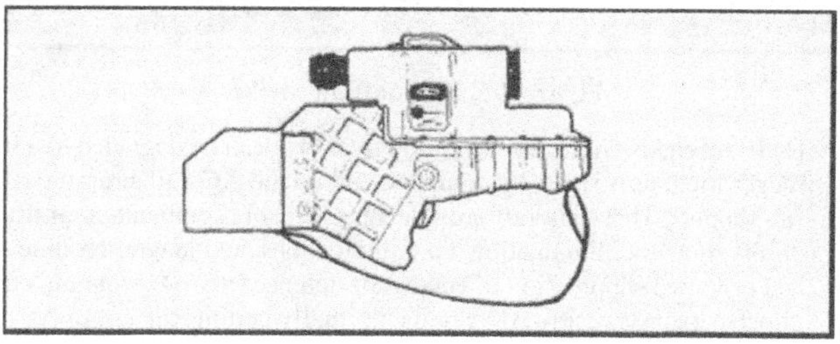

Figure 7-4. AN/PAQ-1 laser target designator.

e. **Aerial fires.**

(1) Aerial fire support is usually the prime means supporting the ranger force due to the distance behind enemy lines at which most ranger operations take place. Aerial fire support can be provided by either fixed-wing aircraft or helicopters.

(a) Fixed-wing aerial fire support may come from USAF, USN, or USMC units. The type of unit providing support, the aircraft, and the mix of ordnance carried all affect the fire support planning and coordination process. Some aircraft have a night and all-weather strike ability enabling them to support the ranger force during any level of visibility. Operations during poor weather that limits visibility to less than 3 nautical miles are still somewhat restricted. The ranger fire support coordinators must ensure that the correct aircraft are requested and employed effectively on the enemy. The

TACP directs and adjusts aerial fires in the objective area.

(b) The ranger force can use ground laser target designators to pinpoint targets for air strikes, as well as electronic navigation aids to permit nonvisual air strikes (beacon bombing). The ranger rifle company FIST or the TACP can control a laser-designated standoff air strike (see Figure 7-5).

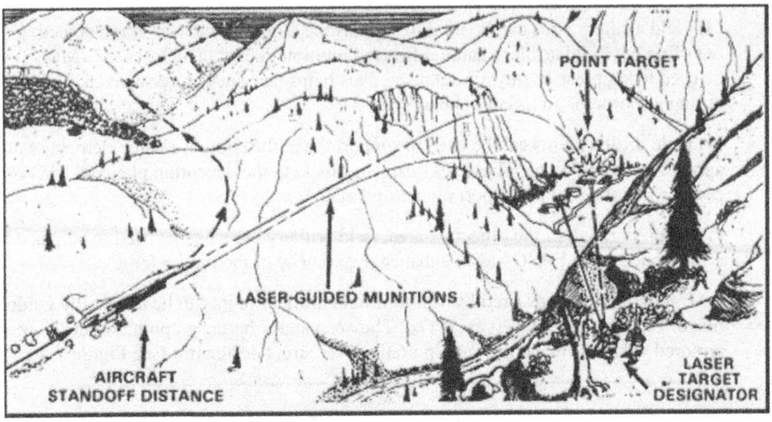

Figure 7-5. Standoff air strike.

(c) If the enemy ADA ability is not great or it can be degraded to a low level, the ranger force uses specially equipped and armed AC-130 aircraft (see Figure 7-6) for fire support. These aircraft provide an invaluable combination of firepower, night observation and illumination, communications, and long loiter time. A well-planned and executed suppression of enemy air defense (SEAD) program, coupled with ECM directed against enemy ADA units, normally permits the use of AC-130 aircraft.

Figure 7-6. AC-130 aircraft.

(2) The attack helicopter, armed with a mix of antitank guided missiles (ATGM), 2.75-inch rockets, 20-mm cannon, and 40mm grenade launchers, is accurate and responsive aerial fire support. The ranger force commander plans to use all sources of aerial fire to help accomplish his mission. However, the nature of ranger operations may preclude the extensive use of armed helicopters due to their limitations. Some attack helicopters are limited in range and lack an all-weather ability. They are restricted during operations at night by a lack of sophisticated night vision devices. They may lack the speed to go with the ranger force on some heliborne insertions. Attack helicopters may be used to escort and assist the ranger force as it crosses the FEBA. They may also be used to conduct feints and demonstrations to cover the insertion of the ranger force. As the MV-22 Osprey and the AH-64 Apache enter service, more use will be made of these aircraft to support ranger operations beyond the FERA, (See Figures 7-7 and 7-8).

Figure 7-7. MV-22 Osprey.

Figure 7-8. AH-64 Apache attack helicopter armed with Hellfire missiles.

(a) If attack helicopters are used to support a ranger operation, planned indirect fires are normally delivered along entry and exit corridors. Attack helicopters approach and depart the objective area using nap-of-the-earth flight profiles.

(b) Fires from armed helicopters are normally requested and controlled by the company FSO or one of his FOs, operating on a special ground-to-air net. The laser target designator may be used to precisely identify targets for the AH-64 Apache. Friendly unit locations may be marked by smoke, panels, lights, mirrors, or infrared sources.

f. **Naval gunfire.**

(1) During amphibious assaults or operations near a coastline, the ranger force may receive indirect fire support from naval gunfire. Normally, a ranger battalion would be supported by either a destroyer or a cruiser in a direct support role. However, the type and importance of the mission, the type of targets, the ships available, the hydrographic conditions, and the enemy capability determine how many and which type of ships are provided to support the ranger force.

(2) Naval gunfire is characterized by large volumes of highly destructive, flat-trajectory fire. Planned strikes in support of the ranger force may also include surface-to-surface missile fire. Some naval guns can fire a laser-guided projectile much like the Copperhead. This projectile, called a Deadeye round, is terminally guided the same as the Copperhead. Deadeyes can be guided either by an attached spotting party from the ANGLICO or by ranger LTD teams.

(3) When ranger units are being supported by naval gunfire, a shore fire control party is attached to the ranger force, This party normally consists of a liaison team and several spotter teams (forward observers). The liaison team is integrated into the operations of the fire support element (FSE) at the ranger battalion TOC. The spotter teams are attached to the ranger rifle companies. The shore fire control party LNO is the ship's representative to the ranger force commander, through the FSO. The spotter teams request and adjust fires from surface vessels. They can also request and control air strikes by carrier-based aircraft.

(4) The LNO and the spotter teams operate in the ground spot net, communicating with the ship by high frequency (HF) radio to request and adjust naval gunfire. The spotter teams communicate with the LNO using very high frequency (VHF) radios. The LNO also can communicate with aircraft using ultrahigh frequency (UHF) radios.

(5) Coordination and control measures that apply to naval gunfire are the same as for field artillery, with the addition of two terms, which are:

(a) Fire support area (FSA)--a sea area within which a ship may position or cruise while firing in support. It is labeled as FSA and numbered by a Roman numeral - for example, FSA VII.

(b) Fire support station (FSS)--a specified position at sea from which a ship must fire. This is very restrictive positioning guidance. It is labeled as FSS with a Roman numeral--for example, FSS VII.

g. **Air defense artillery.**

(1) Army air defense artillery units are rarely used to directly support the ranger force since

ranger operations normally take place deep behind enemy lines. However, it is possible that during the insertion or extraction phase of an operation, ADA units may support the ranger force as it crosses the FEBA.

(2) The ranger regiment's fire coordination element and the FSO in the ranger battalion are mainly concerned with adequate coordination of ADA fires. This prevents friendly units from being engaged. It allows the integration of ADA fires into the scheme of maneuver. The fire support planners ensure that enough coordination has been made so that ADA units are aware of the ranger force and know enough about its flight paths, routes, and altitudes to prevent engagement by friendly units. When conducting this coordination, care must be taken to pass only needed information to the ADA unit. To aid coordination, an ADA officer may be attached to the ranger force.

(3) The plan for a heliborne or airborne insertion and extraction of the ranger force must include the correct identification-friend-or-foe (IFF) codes, the flight profiles, the times of FEBA crossings, and the estimated number of aircraft making the crossing. There must also be a plan for crossing by straggler aircraft and battle-damaged aircraft that cannot reply to IFF interrogation. If possible, the routes selected should avoid concentrations of friendly ADA units.

(4) During an aerial crossing of the FEBA by a ranger force, the regimental headquarters may assign a liaison officer to the headquarters controlling the ADA fires in the crossing area. The LNO is then responsible for close and continual coordination. This ensures that the crossing is made without friendly ADA units engaging the ranger force while it is exiting or entering friendly lines.

(5) During ranger operations, the primary source of air defense fires is shoulder-fired antiaircraft missiles carried by specially trained individual rangers. These weapons provide the ranger force commander a limited defense against attack by enemy aircraft. The decision to take these weapons into the objective area depends on the factors of METT-T. Ranger operations should be completed before the enemy can react with ground forces. However, enemy aircraft may be able to react in time to threaten the ranger force. These lightweight missiles offer a highly effective, short-range counter to enemy air attack.

(6) The ranger force commander must define the weapons control status and rules of engagement for all short-range air defense (SHORAD) teams under his control. Priorities of target engagement must also be determined to prevent multiple engagements of a single enemy aircraft. The ranger battalion FSO and the attached ADA officer are responsible for the coordination of ADA fires within the objective area. Standard ADA weapons-control status and rules of engagement may be modified by the ranger force command.

7-5. Aviation support.

a. **Planning.**

(1) The ranger regiment may need extensive aviation support on some missions. The ranger battalion is organized and trained to move quickly to the objective area by air and to be resupplied by airdrop for short periods.

(2) Often the decision to execute a mission depends on the amount and type of available insertion and extraction aviation support. Normally, the theater commander or JTF commander allocates the type and amount of aviation assets to support a ranger operation. The theater or JTF commander's staff is responsible for the planning and coordination concerning aviation support. The ranger force commander must play an active part to ensure mission success.

(3) Due to the extensive training requirements and specialized operational techniques needed to support special operations, special air operations units in the Army and the USAF have been equipped and trained for the mission. These forces train with the ranger regiment and jointly develop tactics, techniques, and procedures of employment. These special units operate diverse aircraft that have unique capabilities. These help the ranger force commander to effectively employ his units. Command of these special units is maintained at a very high level in the chain of command. They are assigned to support a specific operation. Then they return to their parent unit's control.

b. **Coordination.** In addition to the staff of the theater commander or JTF commander, the coordination of aviation support occurs at all levels in the ranger force. These operations often are conducted in phases. A separate air mission commander is then designated. The AMC is responsible for the completion of each phase of the aviation mission, as well as the close and continuous coordination with the ranger force, commander. The ranger regimental air operations officer and the movements control officer are responsible for the planning, integrating, and controlling of aviation support.

c. **US Air Force.**

(1) The USAF provides special aviation support to the ranger regiment and other elements of the special operations force. The USAF has elements that are equipped with MC-130 Combat Talon aircraft (see Figure 7-9) and AC-130 Spectre gunships, as well as HH-53 Pave Low, UH-60, and UH-1 helicopters. These aircraft are specially equipped to conduct long-range infiltration, exfiltration, attack, and search-and rescue operations. The USAF also provides C-141B and C-130 aircraft for strategic and tactical airlift of the ranger force. There are specially trained USAF crews that can operate using AWADS and SKE. This gives them an excellent means of inserting the ranger force during reduced visibility.

Figure 7-9. MC-130 Combat Talon aircraft.

(2) The AFFOR commander provides the airlift control element (ALCE) support at the departure airfield. The ranger air operations officer ensures that coordination of aircraft load limits, air movement times, specific load plans, and manifests is conducted in an effective and timely manner. The air operations officer and the movements control officer ensure that a departure airfield control officer (DACO) from the deploying unit is designated to provide face-to-face coordination at the departure airfield.

(3) The USAF normally provides an airborne battle command and control craft. This is a specially equipped C-130-type aircraft that provides communications and data system down-links to ground stations.

(4) The terminal guidance needed for assault landings and air traffic control in the objective area is provided by a specially trained and equipped CCT. This element has habitual training and operations relationships with the ranger units. The USAF is also responsible for the specially trained and equipped weather teams that can be sent into the objective area with the CCT.

(5) The USAF Tactical Air Command (TAC) includes fighter, interceptor, ground attack, electronic countermeasure, early warning, reconnaissance, and airborne command and control squadrons, any of which may be used to support the ranger force. The Tactical Air Command provides the results of aerial reconnaissance to the AFFOR, who then processes it for use by the theater or JTF commander's staff. This reconnaissance may be by low-level photo, imaging infrared, side-looking airborne radar (SLAR), low-light-level television, or overhead photography.

(6) The Military Airlift Command (MAC) equips and trains special long-range SAR aircraft and crews. These aircraft, mainly helicopters, are equipped with sophisticated navigation, observation, and communications equipment, as well as extended-range fuel tanks and in-flight refueling capabilities (see Figure 7-10). The ranger regiment uses these aircraft for certain missions.

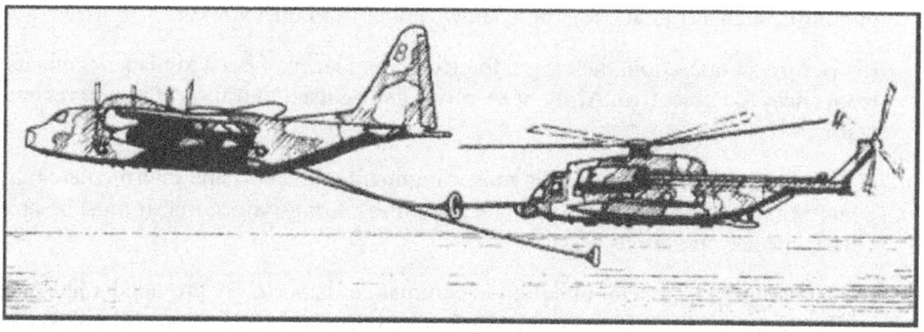

Figure 7-10. In-flight refueling.

d. **US Navy and US Marine Corps.** Both the USN and the USMC have units equipped and trained to support ground forces. The type of aviation support normally provided to the ranger regiment is close air support against targets in the objective area, or as suppression missions against enemy air defense installations. Long-range helicopters from the USMC may be used to support an amphibious operation or to insert special teams during ranger operations. The USN also has a search-and-rescue ability that may be used to support the ranger regiment.

e. **Special operations aviation units.**

(1) The US Army has specially equipped and trained SOA units tasked to support the ranger regiment, as well as other elements of the special operations force. These SOA units have a habitual training association with the ranger regiment and use dedicated aviation assets to -

(a) Insert, extract, and resupply the special operations force.

(b) Conduct armed escort, reconnaissance, surveillance, and electronic warfare in support of the special operations force missions.

(c) Provide airborne command, control, and communications for special operations force elements.

(d) Provide general support aviation during peacetime and contingency operations.

(2) Penetration of enemy airspace is normally done during limited visibility at very low altitudes. Penetrations are planned to maximize the use of existing terrain and exploit blind spots in enemy air defenses. On-board navigational aids are crucial to mission success. There normally is not any terminal guidance to the selected LZ or DZ. Deception operations of a tactical or operational nature help SOA units penetrate to the objective and reenter friendly lines. Advanced helicopters and short-takeoff-and-landing aircraft provide the help needed to do these missions.

(3) To survive enemy air defense systems and prevent mission compromise, SOA units avoid detection. Detected aircraft may reveal the location of supported special operations force ground forces or may be destroyed by enemy ADA fires. The SOA normally avoids detection by operating at night without illumination. This condition requires flight using sophisticated night vision devices. The SOA also operates in other conditions of reduced

visibility, such as clouds, fog, rain, snow, smoke, and dust storms.

(4) Before an operation, the ranger force commander may need to place reconnaissance teams near the objective. Army SOA assets can be used to transport ranger reconnaissance teams.

(5) Elements of SOA provide the ranger regiment with short and intermediate range penetration, resupply, and extraction capabilities. Long-range aircraft must be provided from USAF special operations assets.

(6) The ranger regimental or battalion commander is normally provided a liaison aircraft while the ranger force is in the REMAB. This ensures that he can travel quickly to the dispersed headquarters of the agencies providing support for the mission. This aircraft and crew are provided by the theater or JTF commander from EAC general support aviation assets. It may be a helicopter or fixed-wing aircraft, depending on the location of the REMAB to a suitable landing field.

(7) Army helicopters and fixed-wing aircraft may be used to provide SEMA aerial reconnaissance; radar, infrared and photographic imagery; and electronic intelligence support to the ranger force commander. This support is normally provided by the aerial exploitation battalion of the theater Army military intelligence group.

7-6. Engineer support.

a. The ranger regimental and battalion operations officers plan and request engineer support for ranger operations.

b. The most common engineer-type tasks needed during a ranger operation are the destruction of enemy facilities, equipment, and material by use of conventional explosives; and the interdiction of major lines of communication. To reduce the need for external support, the ranger regiment trains selected individuals in advanced demolition techniques. All members of the ranger regiment are well trained in the basic employment of conventional demolitions. Therefore, there is no need to attach combat engineers to conduct normal demolition missions. If specialized skill or knowledge is needed for a specific mission, qualified personnel can be attached.

c. During the occupation of the REMAB and the rehearsal for an operation, the ranger regiment needs construction engineer support to build simulated target areas. The simulated target areas should be camouflaged to prevent the enemy from recognizing the target. The construction engineers augment the ranger unit's self-help construction efforts. Limited-facilities construction and improvement may also be needed at the REMAB to provide security and isolation. The theater commander or JTF commander provides the construction engineer support.

d. Certain missions, such as those of interdiction, may need engineer special weapons teams. The teams are provided by the theater commander or JTF commander, or they may come from other special operations force assets. They are attached to the ranger force for a specific mission and released at the completion of the extraction and debriefing phase.

e. Some operations may need the ranger force to use captured enemy heavy construction equipment. Selected members of the ranger regiment are trained to operate such equipment.

However, construction engineer personnel may be attached to the ranger force for this mission.

CHAPTER 8

Combat Service Support

Combat service support consists of the logistical and administrative effort to maintain the battalion's ability to fight. The commander forecasts his needs for CSS. He employs CSS units to ensure accomplishment of the mission.

8-1. Ranger support element.

a. The ranger support element is a non-TOE organization formed from elements of CSS units collocated with CONUS-based ranger units augmented by selected personnel from the host installation. The mission of the RSE is to provide all the CSS needed to outload and deploy a ranger unit from CONUS. The personnel designated as part of the RSE have the same alert notification, reaction, and security requirements as members of the ranger unit they support. The RSE is normally built around a command and control group provided by a major CSS unit on the ranger unit's host installation. The remaining CSS elements and teams are then designated. An RSE consists of--

(1) An emergency operations center liaison group that provides timely coordination between the ranger unit and the functional elements of the RSE.

(2) A command and control group that coordinates and directs the actions of the functional elements.

(3) A supply section responsible for the timely issue of all Class II, III, IV, and VII items. This section maintains a stockage of contingency items marked for use by the resident ranger unit. It also issues items out of common stocks.

(4) A transportation section moves equipment and personnel from the unit area to the departure airfield and picks up equipment from local storage sites. It may also operate material-handling equipment and trailers to support loading of aircraft.

(5) An ammunition section responsible for the requisition, storage, issue, loading, and security of the ranger unit's basic load(s) of conventional munitions.

(6) A mess section provides all Class I support to the ranger unit. This includes operating a dining facility, providing food to remote sites, and drawing and issuing operational rations.

(7) A load team loads and unloads all equipment and supplies for the ranger unit. This team prepares pallets for loading, documents hazardous or dangerous cargo, and loads equipment pallets on USAF aircraft. It can also provide limited security for loads.

(8) A communications and electronics section provides secure communications among the RSE subelements, ranger unit, EOC, and DACG.

(9) An engineer platoon that is responsible for the construction of targets and rehearsal areas. This element of the RSE may or may not be activated, depending on the mission given the ranger unit. If the ranger unit deploys from CONUS to the objective area, this platoon has a key role to play in the premission preparation.

(10) A maintenance team whose primary mission is to ensure that the vehicles used by the RSE remain operational. It also provides predeployment maintenance and Class IX support for ranger unit vehicles and equipment.

(11) A medical team that provides unit-level medical care while the organic ranger medical section completes its outloading preparations. This team also provides any Class VIII supplies that the ranger unit may need.

(12) A security team responsible for security at the EOC, ranger unit location, and departure airfield. This team enforces stringent physical security precautions and maintains a constant counterintelligence (CI) surveillance of the outloading procedure. The security team provides armed guards for the protection of ammunition convoys, equipment movement, and vehicles.

(13) A rigger detachment that maintains, packs, rotates, and issues the personnel and equipment parachutes required by the ranger unit. It maintains enough stock of personnel parachutes to conduct a battalion-level parachute assault. The riggers lend technical skill to the rigging and aerial delivery of bulky, unusual, or delicate equipment.

b. The primary mission of the RSE is to provide the support necessary to quickly load and deploy the ranger units. The theater commander provides CSS to the ranger force once it has arrived at the OCONUS REMAB or ISB. The transition of support from the RSE to the CSS assets of the theater commander may be delayed for certain subelements in order to ensure a smooth deployment and enhance mission accomplishment. These subelements are released when the in-theater CSS assets can provide the support. Retained subelements of the RSE may be released from support requirements all at once, released in increments, directed to continue to provide support until mission completion, or augmented by in-theater CSS assets. The maintenance, medical, security, and rigger elements are those most likely to continue providing support OCONUS. This is likely if deployment is to an austere ISB pending immediate employment.

c. Once a ranger unit has been deployed OCONUS and has closed to a REMAB or ISB, the theater commander's staff coordinates CSS. If 1st SOCOM elements have deployed to the theater or operational area, the special operations support battalion (SOSB) of the 1st SOCOM coordinates CSS. Ranger logisticians at regiment and battalion level are mainly planners with limited logistics operations ability. Ranger logistical and personnel officers coordinate with CSS elements who provide such services.

d. The Theater Army Materiel Management Center (TAMMC) is the point of contact for information concerning in-theater logistical assets. In a developed theater of operations, a series of theater Army support groups will have been set up. One of these groups is assigned the mission of supporting the ranger force. The 1st SOCOM may also deploy certain assets available to provide CSS to the ranger force. The many CSS sources, the rapidly changing nature of SOF operations, and the unique CSS requirements of ranger units combine to make providing logistical and personnel support a challenge. The ranger unit's logistics officer must be included in the early phases of planning and deployment.

8-2. Logistics and personnel.

a. The echelons-above-corps CSS structure provides support to the ranger regiment as directed by the theater Army (TA) commander. In a low-intensity contingency involving a corps or less, there may be no echelons-above-corps logistics and personnel support elements. The corps support command (COSCOM) involved may be able to meet all needs from its own sources. However, in a combined operation in a high-intensity environment, the theater commander assembles a large CSS organization to provide support to the TA. This CSS base normally provides for all the ranger unit's requirements. During contingency operations in an area where there is no TA, the JTF commander provides the CSS elements. Until the buildup of these elements is complete, the ranger regiment must be prepared to operate from austere facilities with only the supplies it brings from CONUS.

b. The ranger regiment is normally assigned to the theater command for CSS. The TA commander, as directed by the theater commander, provides CSS to the ranger regiment and sets up support priorities. The EAC CSS organizations are under the command of the TA commander and consist of subordinate area and functional commands. The TA plans, coordinates, and executes its CSS functions through area-oriented support commands and specialized subordinate organizations. During the early stages of conflict, the TA may be organized as shown in Figure 8-1.

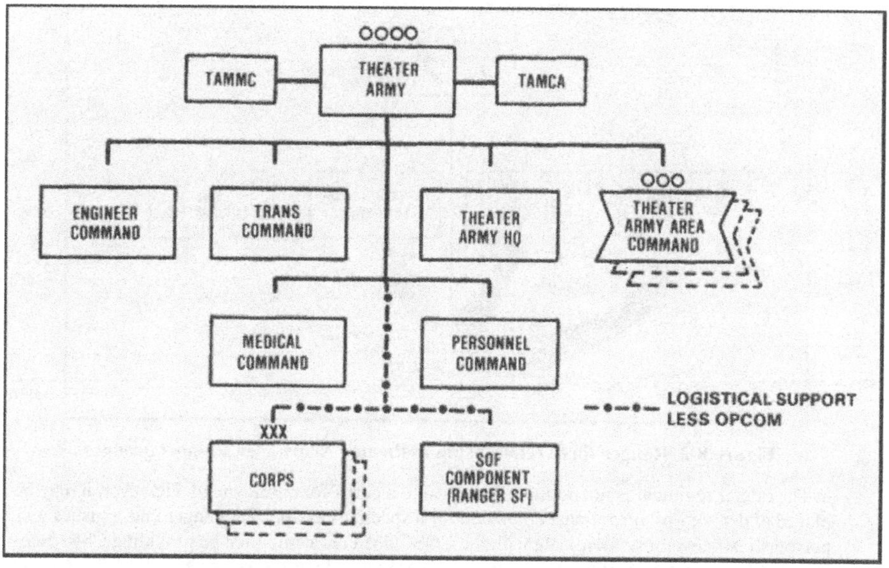

Figure 8-1. Ranger force relationship to theater Army.

c. The Theater Army Area Command (TAACOM) is the key CSS operator in the communications zone. It functions as a major subordinate command under the theater Army. The primary mission of the TAACOM is to provide direct support CSS to units such as the ranger regiment located in or passing through its assigned area. While located in an OCONUS REMAB, the ranger regiment receives most of its logistical support from the TAACOM.

d. The TAACOM area support group (ASG), which is subordinate to the TAACOM, is structured with operating units to provide direct support CSS to units located in or passing through its assigned geographical area of responsibility (see Figure 8-2). Area support groups are usually located along the lines of communication. The ASG is the prime source of CSS for the ranger regiment. It contains the units that actually provide services requested by the ranger unit logistics officer.

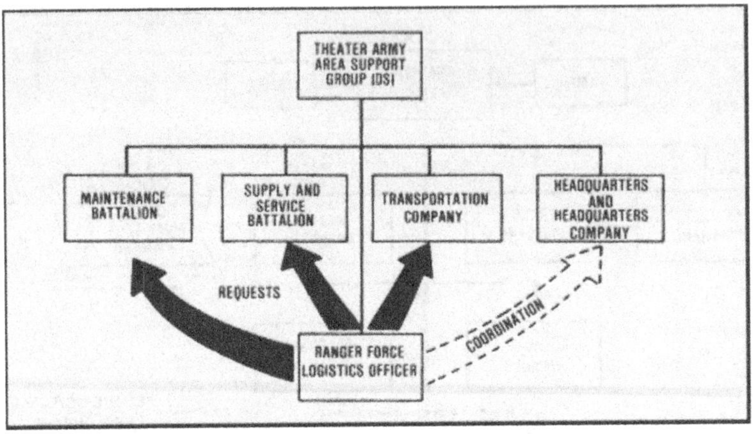

Figure 8-2. Ranger force relationship to theater Army area support group.

e. The ranger regiment is not normally assigned to a corps for employment. However, it may be placed under a corps' operational command for a specific mission. The ranger unit logistics and personnel officers must coordinate with the COSCOM that would then be providing CSS. (See Figure 8-3.) Corps support groups are major subordinate elements of the COSCOM. They provide command and control of COSCOM companies and battalions, which provide direct support (DS) and general support (GS) supply and field services to nondivisional units, such as the ranger battalion. (See Figure 8-4.)

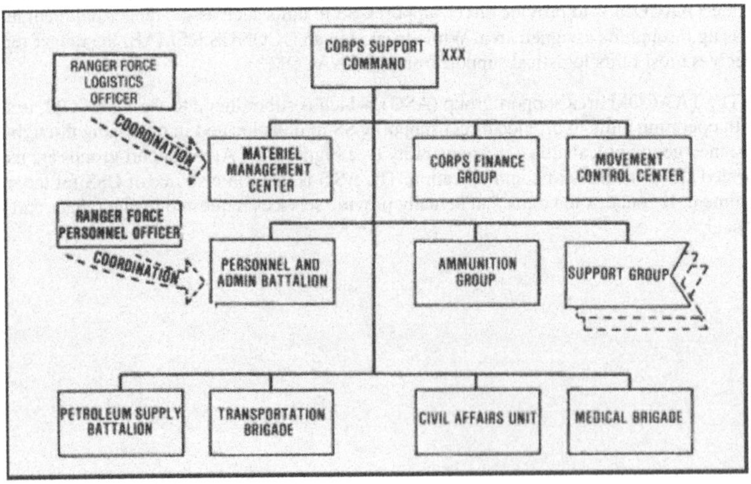

Figure 8-3. Ranger force relationship to corps support command.

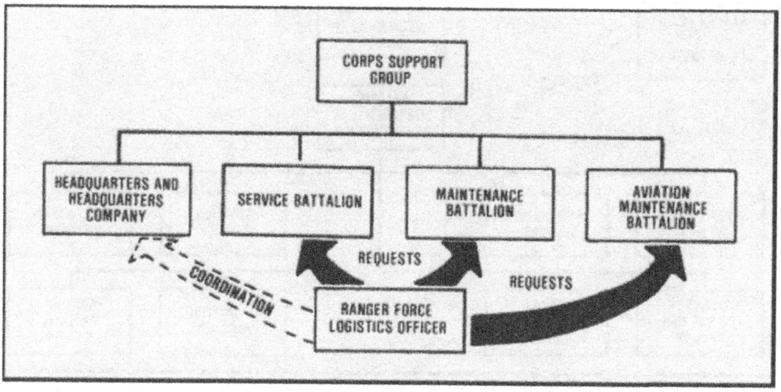

Figure 8-4. Ranger force relationship to corps support group.

f. Normally, the ranger regiment receives its logistical support from the corps support group responsible for the geographical area in which the REMAB or ISB is located. The corps support group provides all supplies except Classes V and VIII. The medical brigade provides Class VIII supply and medical maintenance support. The ammunition group provides Class V support. Financial and personnel services are provided by the corps finance group and the personnel and administration battalion, respectively.

g. Transportation needs are met by the transportation brigade. The ranger unit places ground transportation requests with either the Movement Control Center (MCC) or a designated movement control team. All airlift supply requirements are passed IAW FM 100-27. Standard and nonstandard supplies are normally requested from the supporting direct support unit (DSU). If authorized by the corps commander, the ranger regiment can place requirements for nonstandard items on the appropriate division within the corps MMC.

h. The COSCOM MMC (see Figure 8-5) performs supply and maintenance management for all classes of supply. The ranger unit logistics officer may coordinate with the MMC division chiefs or operations section.

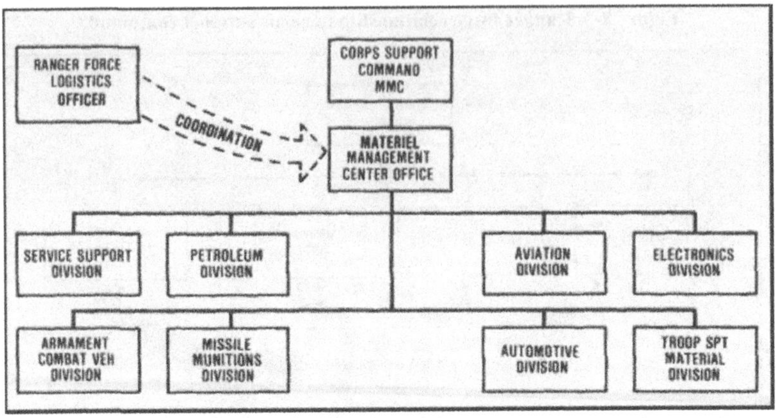

Figure 8-5. Ranger force relationship to COSCOM MMC.

i. Critical logistics operations, such as those supporting ranger and other SOF elements, need intensive management. They may be assigned to special staff sections that act as special operations force logistics managers. If designated, a logistics manager for special operations force elements would be a major point of contact for logistical support for the ranger regiment.

j. The ranger regiment may also receive logistical support from the SOSB. Normally, the SOSB provides administrative and logistical support to the Headquarters, Army Special Operations Forces (ARSOF). It complements the support provided by the Army component in the theater of operations by coordinating, augmenting, or assisting CSS forces. The SOSB provides unique quick-reaction support to the ARSOF or deploying SOF elements.

k. The ranger regiment does not make inordinate demands on the logistical system. However, a ranger unit may need certain nonstandard or low-density equipment. The concurrent need for nonstandard Class IX items requires management by logisticians to ensure support.

(1) Ranger battalion mess sections are responsible for Class I operations at the REMAB. The ranger regimental headquarters, which has no mess section, either collocates with one of the battalions or is supported by a mess facility operated by a nearby unit. It may share this facility on an area basis. The ranger unit personnel officer coordinates with the supporting DSU, which initiates the issue of Class I supplies.

(2) Class II, IV, and IX support to the ranger force is normally not of a great volume. It may consist of some nonstandard items. Depending on the need, the item, and the security, the SOSB may get some nonstandard items directly.

(3) Class III support needed by the ranger regiment is small and limited to standard items. The normal source of supply for Class III is the supply and service company providing area support.

(4) Class V support needed by the ranger regiment is small. It may include requests for nonstandard items of ammunition and explosives. The ranger logistics officer passes requests for standard items to the ordnance company (conventional ammunition) of the ammunition battalion assigned to the TAACOM ammunition group. Requests for nonstocked items are passed through the TAACOM Materiel Management Center or 1st SOCOM channels. The ranger unit must contact the ammunition company early to ensure a continuous supply of ammunition.

(5) Class VII supplies needed for the ranger regiment are often nonstandard or in excess of TOE. (See Appendix C for an equipment density listing.) All nonstandard or excess requests must be coordinated through the TAMMC or 1st SOCOM channels.

(6) Class VIII supplies for the ranger regiment are normally standard items supplied by the area support Class VIII point. The ranger regiment personnel officer sets up the Class VIII supply points of contact through the ranger battalion's medical officer.

(7) Aerial delivery equipment is provided by the airdrop equipment repair and supply company of the theater Army support group (TASG) (GS). There is normally one such company for each TAACOM. This unit supplies and repairs aerial delivery equipment. It does not pack and rig supplies for aerial delivery. The ranger regiment's requirements for aerial delivery supplies normally consist of standard items. Certain low-density items, such as HALO or HAHO parachute equipment, may also be provided through 1st SOCOM channels.

(8) The supporting motor transportation group of the transportation command is responsible for unit movement of the ranger regiment. The ranger regiment does not have any organic transportation assets. It depends on outside support whenever motor movement is needed.

8-3. Resupply.

a. Resupply operations for ranger units are normally covered during the planning and coordination phase of the mission. Ranger units try to complete their mission quickly and then be extracted or exfiltrate from the objective area. Ranger units carry into the objective area the equipment and munitions they need. They do not normally remain long enough to need extensive resupply. Certain operations may require the use of bulky or heavy items of equipment that cannot be carried by personnel. If the ranger force is airlanded on or near the objective, one or more of the aircraft may contain palletized supplies. These can be quickly off-loaded for future use.

b. If the ranger force is inserted into the objective area by parachute assault, aircraft can also drop resupply loads either just before or just after the personnel drop. These loads might include extra water, ammunition, demolitions, or barrier material. During peacetime, preloaded airdrop pallets are stored at CONUS depots for such contingency use. These mixed loads can be quickly moved to a departure airfield and inserted along with, or following, the ranger force. By dropping a resupply of water, food, equipment, and munitions with the ranger force insertion, the heavy-drop aircraft do not have to conduct a separate penetration of the enemy air defense zone. The ECM and SEAD programs used in the insertion of the personnel aircraft mask the penetration of the heavy-drop aircraft.

c. The aircraft dropping ranger personnel can also carry door bundles to deliver bulky or heavy items to the objective area. These door bundles, normally made up of A-7 and A-21 containers, may be color-coded to identify different loads or units. The most common door bundle loads are such items as 90-mm recoilless rifles and ammunition, 60-mm mortar ammunition, Dragon missiles, SHORAD missiles, or more demolition material. These bundles are pushed from the personnel aircraft just before the lead jumper's exit. They may be marked with lights or infrared sources for locating them at night.

d. If the tactical situation requires that a ranger force receive a resupply during the course of an operation, the most common method would be by airdrop. Airdrop could be needed if the operation takes longer than planned, initial supplies are lost, or changes occur in the operational plan. The ranger unit logistics officer and operations officers normally plan for resupply needs and schedule follow up loads of Classes I, II, IV, V, and VIII. There are two different airdrop requests channels for the ranger force to use:

(1) Nonemergency airdrop requests are sent through logistics channels. The ranger LOC sends requests for the airdrop of supplies and equipment to its supporting supply unit, and the supporting unit forwards the requests to the COSCOM or TAACOM MMC. The MMC coordinates with the movement control center (MCC). Jointly they direct the actions of the supply and transportation units. The MCC sends the request to the joint transportation board (JTB). The board validates it and assigns it a priority. The JTB then sends the request to the airlift control center for action. If the items requested are not on hand in the airdrop company, the MMC directs the proper supply activity to prepare the items for shipment to the airdrop unit. The MMC coordinates with the MCC, and the MCC ships the items. The airdrop unit prepares the supplies for the mode of delivery to be used. It also makes sure the supplies are delivered to the air terminal at the right time. (See Figure 8-6.)

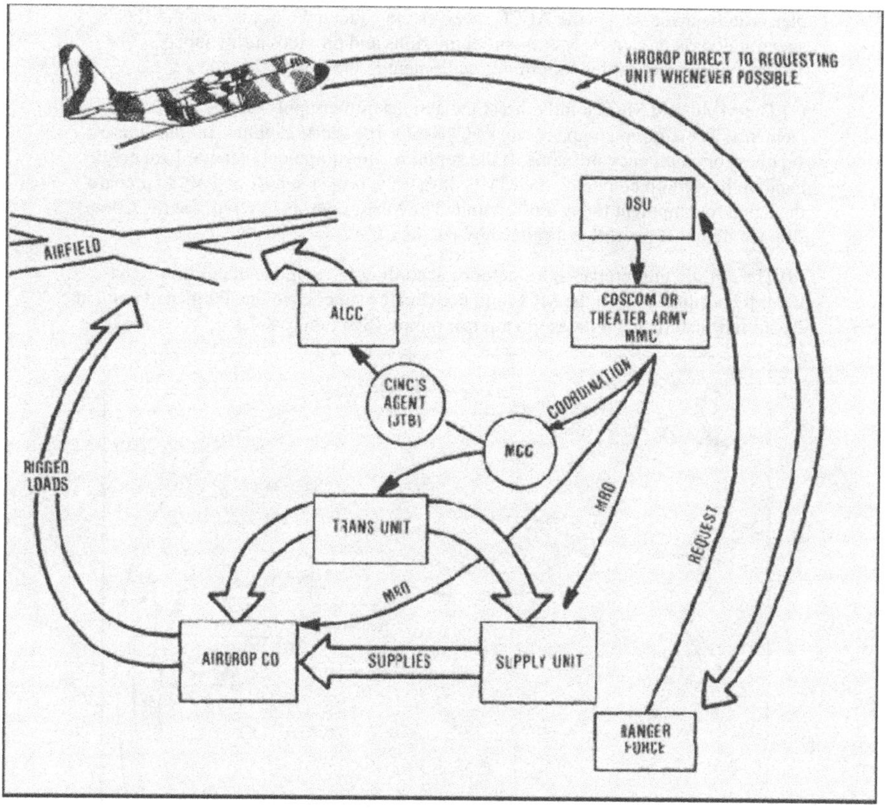

Figure 8-6. Ranger requests for nonemergency airdrop.

(2) Emergency requests for airdrop are passed through command channels in the quickest, most secure way.

(a) The request is usually sent to the command level that has authority to approve it, usually the tactical operations center at corps or above. The Air Force airlift officer at the direct air support center (DASC) coordinates with the Assistant Chief of Staff, Operations (G3) at the tactical operations center to determine whether airlift resources can be used for the mission.

(b) The TOC sends a request for an immediate airlift to the joint force commander's (JFC) designated agent for validation. At the same time, the airlift liaison officer advises the airlift control center (ALCC) that a request is coming through channels so plans can be made. When the ALCC receives the validated request from the JFC agent, it diverts or cancels less essential missions and provides airlift support. The TOC sends the request for supplies or equipment to the MMC.

(c) The MMC and MCC jointly direct the actions of the supply and transportation elements. The airdrop company stocks Classes I, III, and V supplies. It holds these supplies for emergency missions. If the supplies and equipment requested are not on hand in the airdrop company, the MMC directs the proper supply activity to prepare the items for shipment to the airdrop unit. The MMC coordinates with the MCC to ship the items. The airlift is arranged by the ALCC.

(d) The airdrop unit prepares the supplies according to the mode of delivery to be used. It coordinates with the MCC and delivers the supplies to the designated point at the air terminal (usually the cargo aircraft ramp). (See Figure 8-7.)

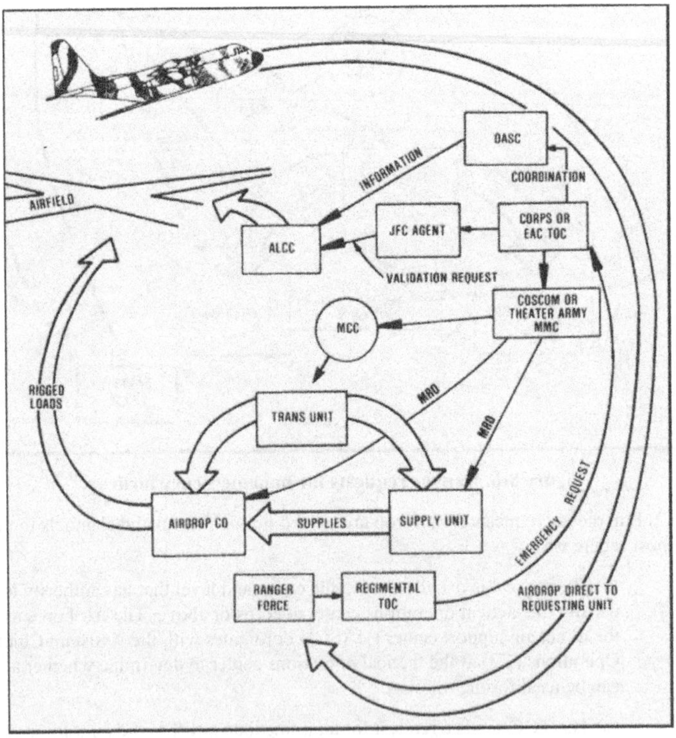

Figure 8-7. Request channels for emergency airdrop.

e. Aircraft delivering airdrop resupply deep behind enemy lines need to take precautions to avoid enemy detection and antiaircraft fire. The USAF is responsible for most of these precautions, such as ECM, suppression of enemy air defense (SEAD), and WILD WEASAL support. These do not concern the ranger commander. The choice of the type of parachute delivery system to use, however, may affect the ground commander's plan. The ranger force commander must specify any tactical restrictions that would require a specific type of delivery system. The ranger LOC and regimental logistics officer must advise the ranger force commander on the type of parachute system to be used. Conflicts must be quickly resolved so as not to interfere with the tactical operations.

(1) The safest way for the airdrop aircraft to penetrate enemy air defense and remain undetected is often by use of very low-level flight. There are four low-level parachute delivery systems that can be used without the aircraft being required to climb to a higher drop altitude and risk detection by enemy ADA.

(a) The low-altitude parachute extraction system (see Figure 8-8) needs a length of level ground with enough clearance for the delivery aircraft to make a very low pass over the extraction zone. The load is extracted by parachute from the aircraft and slides to a stop on a rugged, skid-type pallet. Heavy loads can be delivered using this system.

Figure 8-8. Low-altitude parachute extraction system.

(a) The containerized delivery system (CDS) (see Figure 8-9) consists of multiple, individually rigged A-22 containers, each with its own parachute, weighing up to 2,000 pounds each. The system provides single-pass delivery of up to 16 containers by C-130 aircraft and up to 28 containers by C-141 aircraft. The CDS loads may be delivered into drop zones using multiple points of impact to allow for tactical separation. The ranger force commander can direct the use of multiple points of impact, noting the advantages to be gained against the requirement for multiple passes.

Figure 8-9. Containerized delivery system.

(c) The high-speed, low-level, airdrop system (HSLLADS) (see Figure 8-10) is a single A-21 container specially rigged to withstand the parachute opening shock when airdropped from C-130 aircraft at high speed. This system can be used to deliver up to 600 pounds for each container, with a maximum of four containers for each pass, at speeds up to 250 knots.

Figure 8-10. High-speed, low-level, airdrop system.

(d) The CTU-2/A is a high-speed, aerial-delivery container (see Figure 8-11) that can be used to deliver supplies from high-performance aircraft flying at a minimum altitude of 300 feet above ground level and a maximum airspeed of 425 knots. The CTU-2/A is carried on the bomb racks of fighter or bomber-type aircraft. Upon release, a pilot parachute deploys the main chute and the container descends slowly.

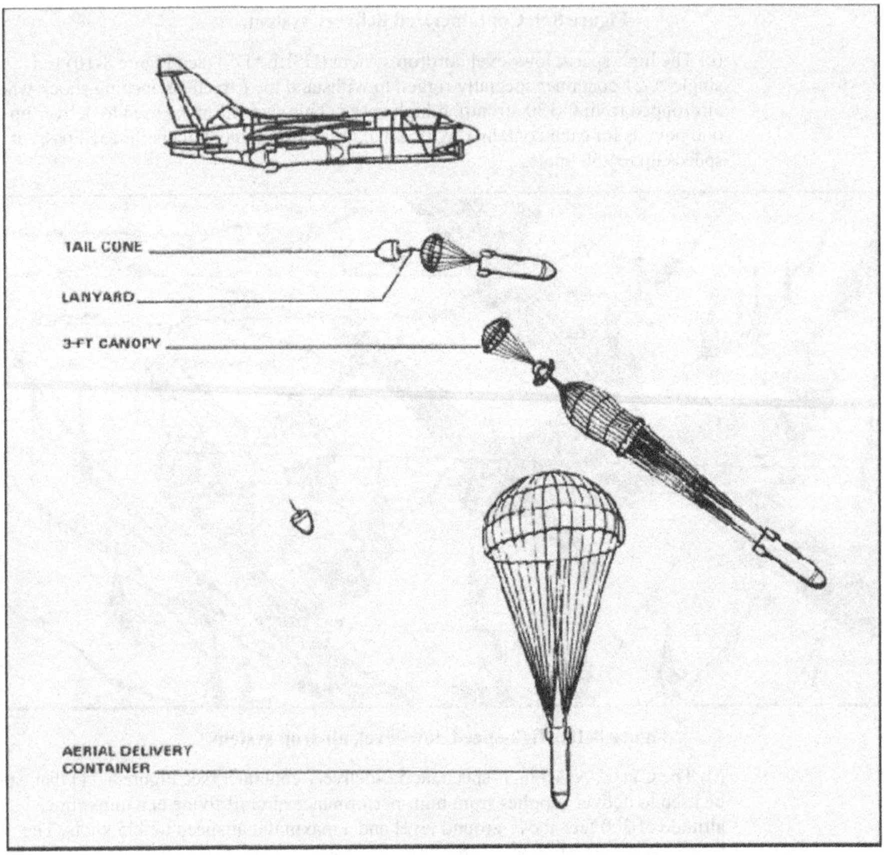

Figure 8-11. CTU-2/A high-speed, aerial-delivery container.

The CTU-2/A can be used to deliver up to 500 pounds of supplies such as weapons, water, food, or munitions. The container can be destroyed by burning. The main advantage of this system is that it can be delivered by high-performance aircraft deep behind enemy lines and in a dense air defense environment. The accuracy of this system is equal to that of a conventional bomb strike.

(2) Sometimes it may be better to drop resupply loads to the ranger force from a high altitude. The HAARS (see Figure 8-12) permits containerized unit loads weighing from 200 to 2,000 pounds to be delivered from aircraft at speeds up to 150 knots from up to 25,000 feet AGL. The HAARS consists of a cargo parachute, an airdrop container, an altitude

sensor, and a pilot chute. The pilot chute gives the descending bundle a speed slightly greater than an accompanying HALO parachutist. The system provides for the steady free-fall descent of loads from altitudes between 2,000 and 25,000 feet to an altitude at which a barometric sensor actuates deployment of the main parachute. This allows safe and accurate delivery of loads onto unprepared drop zones. The HAARS can deliver a payload to within 260 meters from a target impact point from a 10,000-foot altitude, with a proportional degree of accuracy from 25,000 feet. Ranger units can use the system for resupply of battalions and smaller units with rations, ammunitions, and medical supplies, breaking down the containerized material into man-packed loads.

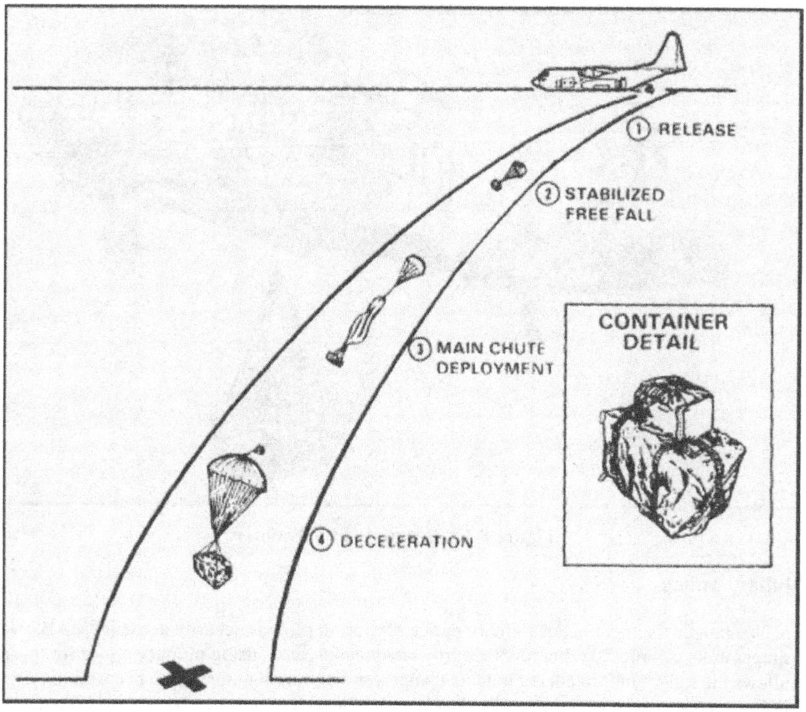

Figure 8-12. High-altitude, airdrop resupply system.

(3) The ranger force may need aircraft loads delivered during periods of poor visibility. The adverse weather aerial delivery system allows aircraft to drop resupply bundles accurately in adverse weather or darkness. These AWADS operations can be conducted safely and effectively with a minimum 300-foot ceiling AGL and a minimum visibility of less than half a mile. The minimum-drop altitude is 500 feet above the highest obstruction. These minimum ceiling or visibility restrictions do not apply to the Combat Talon aircraft.

f. Resupply to the ranger force can be provided by Army helicopters (see Figure 8-13). Normally this is done by SOA units. The danger to helicopters during long-range, low-level penetration of enemy rear areas requires that resupply be conducted when there is limited visibility.

Figure 8-13. Resupply bt helicopter.

8-4. Military police.

a. The ranger regiment needs military police support to provide security at the REMAB. Although ranger units can provide their own security and local defense, using military police for this purpose allows the ranger unit to concentrate its manpower and training on mission preparation. A military police security company is normally designated to provide support to 1st SOCOM elements located in the communications zone. Military police provide security patrols of the REMAB, convoy security, and vehicular search. They provide entry and exit control of the sensitive compartmented information facility (SCIF).

b. A preemptive strike by enemy special operations forces is a threat to the ranger regiment while located in the REMAB; either by conventional attack or by terrorist-type bombing. The preparation and location of the headquarters, the type of buildings, the location of landing zones, the traffic flow, and the size of the local population all influence the size of the military police unit. A military police platoon, augmented with bomb and guard dog teams, should be enough security at the REMAB. These security forces would not normally deploy from the REMAB to an ISB with the ranger unit.

8-5. Counterintelligence.

Counterintelligence support to the ranger regiment is normally provided by a CI team from the military intelligence group at TA level. The CI team provides a liaison and informational flow from the CI agencies operating within the theater. It also acts as advisor to the ranger unit commander on OPSEC while performing CI surveillance operations in the vicinity of the REMAB. The CI support may also be provided through 1st SOCOM channels. It is through 1st SOCOM that liaison with national intelligence and counterintelligence agencies is achieved.

8-6. Personnel service support.

The personnel service support needs of a ranger unit are much like that of any other unit of the same size. To receive the support his unit needs, the ranger personnel officer must meet with the proper agencies within the theater of operations.

a. **Strength accounting.** Within the ranger regiment, strength accounting is conducted through normal personnel channels. Regimental strength figures are passed quickly to the TAACOM or COSCOM. These figures are crucial to the theater commander for planning future operations. When matched with casualty reports, they provide the information needed for replacements.

b. **Replacement operations.** Ranger units need exceptionally well-trained and physically conditioned personnel as replacements. Ranger operations require that replacements train with the unit before they can become an effective member of the team. There are five sources of replacements for ranger units:

(1) **In-theater volunteers.** Use of in-theater volunteers, preferably with experience in ranger units, is the quickest way to replace personnel. These volunteers have a knowledge of the basic skills required of rangers and have some operational experience. The ranger regimental personnel officer coordinates with the theater personnel officer to screen and select those volunteers with the required MOSs and grades. Volunteers from other special operations force components, or from other theaters of war, are identified by DA and screened for assignment to the ranger regiment. Even though these volunteers have individual training and experience, it takes several weeks to become fully integrated into the ranger unit's operations and to achieve the needed skill level.

(2) **Reorganization of units.** This method of reconstitution is used when two or more ranger units have high personnel losses. The remaining members of these units are reorganized into a single unit. This method provides the ranger regimental commander with enough ready-trained, physically fit personnel to form a full-strength unit. Extensive unit training must be conducted before such a reorganization regains the unity and esprit of its component parts. The high standards found among the subelements of the ranger regiment makes this a good method of reconstituting a ranger unit. The units that have lost all their personnel may be reconstituted later, using some other method, or left unreconstituted,

depending on the situation.

(3) **Hospital returnees.** These personnel, former members of the ranger regiment who were injured, evacuated, treated, and have now recovered, are valuable assets. They provide the unit with a major reconstitution ability. These personnel should be returned quickly to their original battalion. This enhances unity and increases morale and esprit. Returnees may not be physically able to return to full duty right away. While recuperating, they may be assigned to limited duty as instructors. The ranger unit uses the returnee's combat experience and individual skills to screen and train newly assigned volunteers. Lessons learned in combat can help prepare the unit for future operations.

(4) **Replacement volunteers from CONUS.** These personnel have training in the basic skills and may have completed the ranger course, but they do not have any ranger unit experience. They need several weeks of unit training to become prepared for combat operations.

(5) **Ranger indoctrination program.** During wartime, this program provides screening and training of newly assigned personnel. Volunteers who have neither served in a ranger unit before nor have graduated from the ranger course require extensive individual training before they are ready to begin unit training with a ranger battalion. This method of reconstitution of ranger units requires the longest training period. During this reconstitution phase, a ranger battalion needs extensive use of ranges, maneuver areas, and support by aviation and other special operations-type units. The ranger regimental commander may choose to move the program from CONUS to a location near the regiment's REMAB.

c. **Casualty reporting.** Casualty reports are handled through regimental personnel channels and must be timely and accurate. The reports are used to check strength accounting data, to assist graves registration personnel in the recovery and identification of remains, and to notify next of kin.

d. **Health service.** The health service support needed by the ranger units deployed OCONUS is usually small. Area medical support provided by host country or theater medical command assets usually meets their needs. Evacuation of wounded members of the ranger force begins upon extraction from the objective area. Personnel wounded in action are evacuated through USAF or Army channels to the nearest medical facility that can provide treatment. The ranger force medical officer and personnel officer coordinate with the area medical support facility providing preventive and routine medical care to the unit while at the REMAB. Vaccine needed for innoculation of the ranger force members and any special Class VIII supplies needed for the mission are drawn through normal medical channels.

e. **Financial service.** Financial support varies with the situation. In mid- and high-intensity conflicts, regular paydays are suspended in the theater of operations. As a nondivisional unit, the ranger regiment coordinates with the theater finance center or corps finance group for financial support. In the REMAB, ranger units are provided financial support from the closest finance support unit. The ranger unit personnel officer appoints Class A agents to help with the disbursement of funds. The ranger force may be given a specific amount of local currency or military scrip depending on the length of time in the REMAB and the operational situation. The ranger unit personnel officer also coordinates with the theater finance center to get any third-country currency that may be included as part of an escape-and-evasion kit provided to ranger elements.

f. **Postal service.** Postal service to the ranger unit is through normal channels. Mail for members of the deployed force is forwarded by the rear detachment commander to an established theater Army general support postal activity. This activity forwards the mail to the postal unit designated to support the REMAB site. The ranger unit commander appoints subordinate unit officers as censors. The exact location of the REMAB, and all aspects of the mission and its location, remains classified until released by an official Army spokesman. The ranger unit commander must ensure that the forwarding of mail, or the writing home of personal letters, does not endanger the ranger force mission. All outgoing mail is censored during wartime. For security reasons, outgoing mail may be held for a short time before it is forwarded. Mail from the ranger unit is never sent by host country or other nation's postal services.

g. **Chaplain activities.** Chaplain activities are provided by the chaplain assigned to the ranger battalion. He provides unit coverage through services, rites, and sacraments regardless of his or the unit member's religious affiliation.

h. **Legal service.** Ranger units do not need extensive legal service support. Legal service support is obtained through coordination with the theater Army staff judge advocate.

APPENDIX A

Organization and Equipment Introduction of the Ranger Regiment

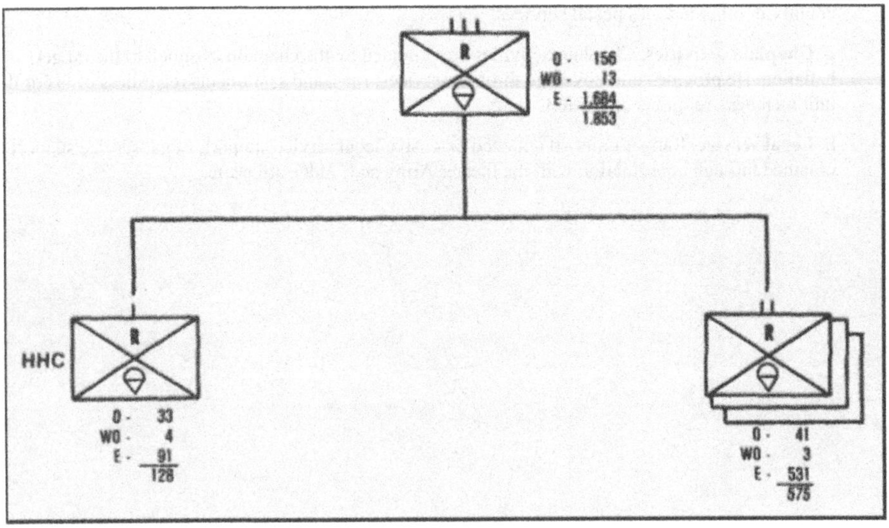

Figure A-1. Ranger regiment.

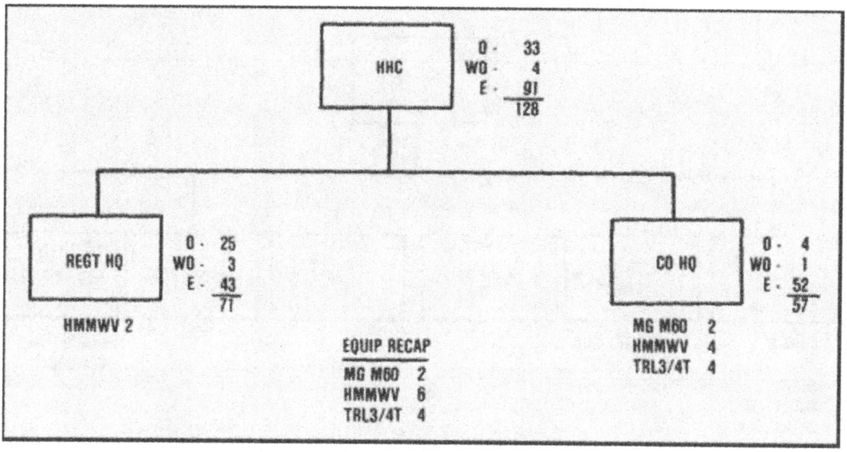

Figure A-2. Ranger regimental HHC.

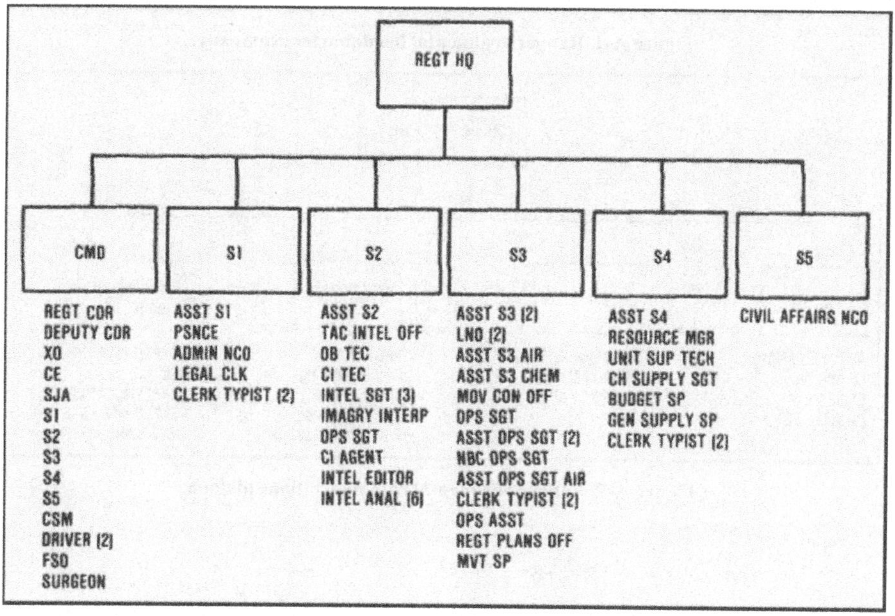

Figure A-3. Ranger regimental staff.

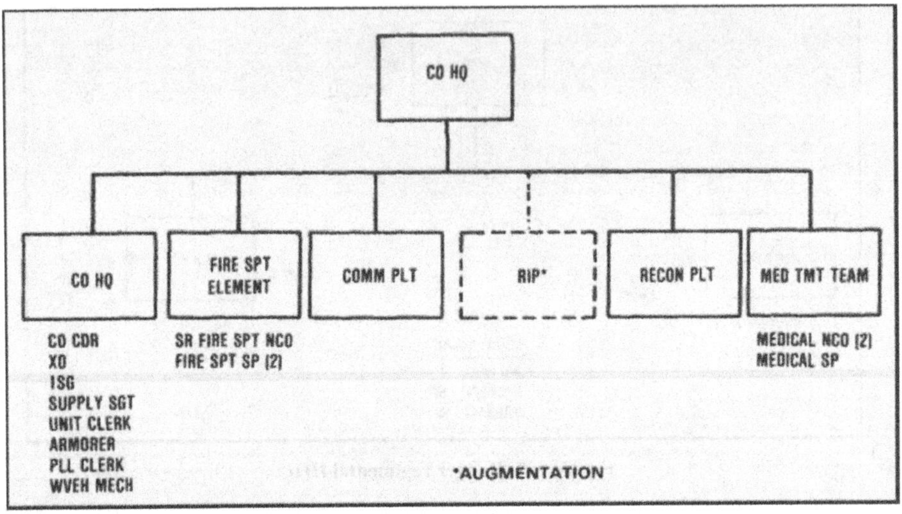

Figure A-4. Ranger regimental headquarters company.

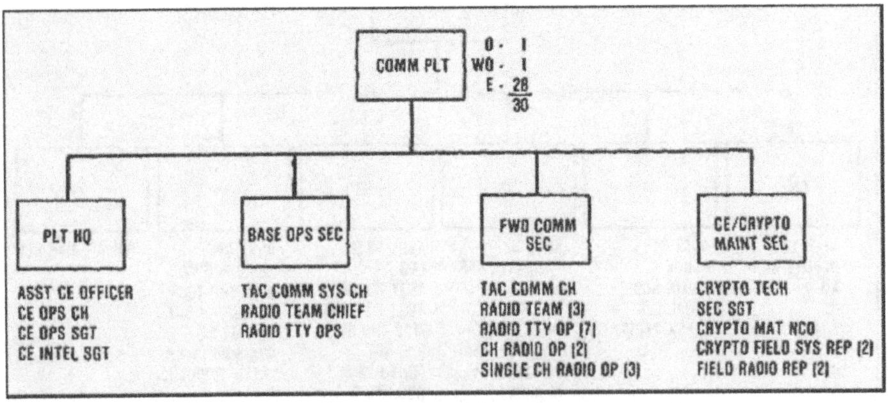

Figure A-5. Ranger regimental communications platoon.

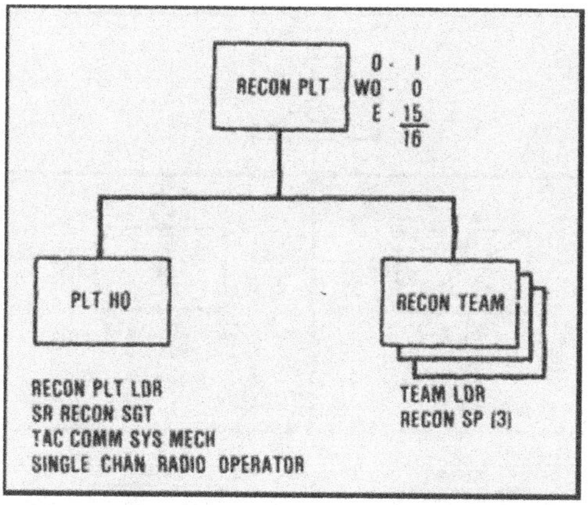

Figure A-6. Ranger regimental reconnaissance platoon.

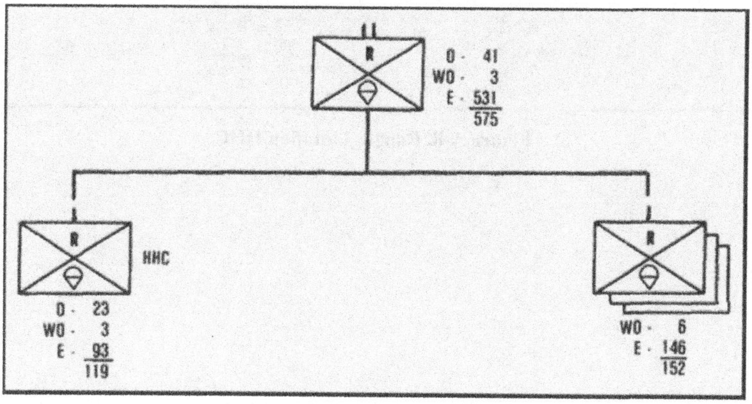

Figure A-7. The ranger battalion.

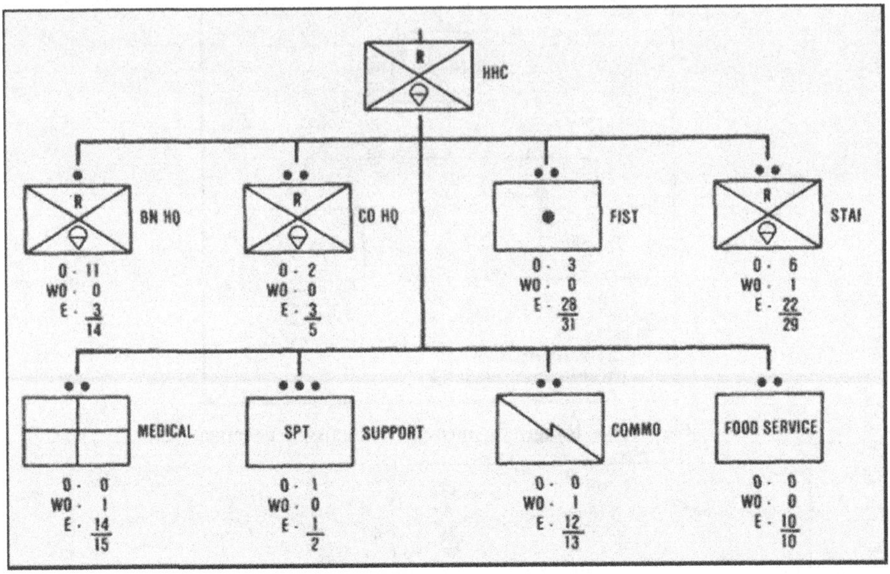

Figure A-8. Ranger battalion HHC.

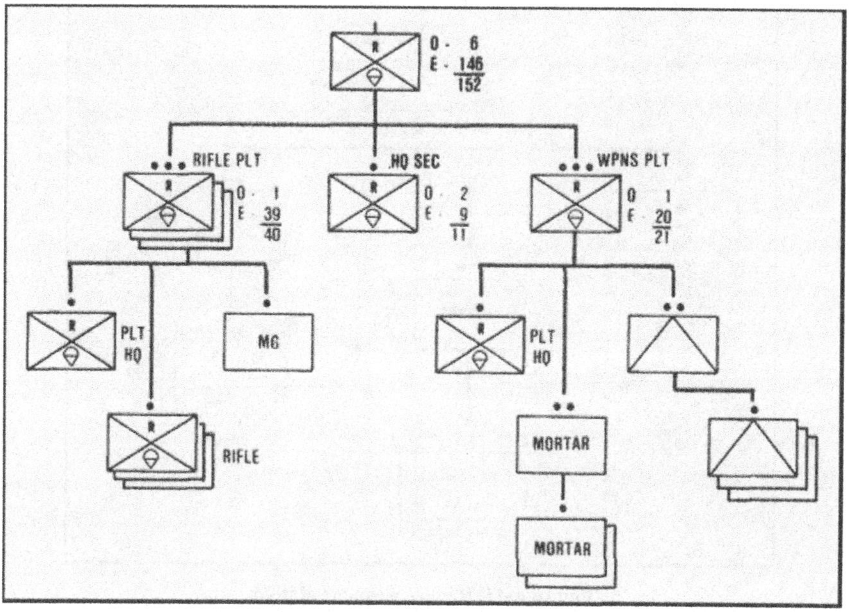

Figure A-9. Ranger rifle company.

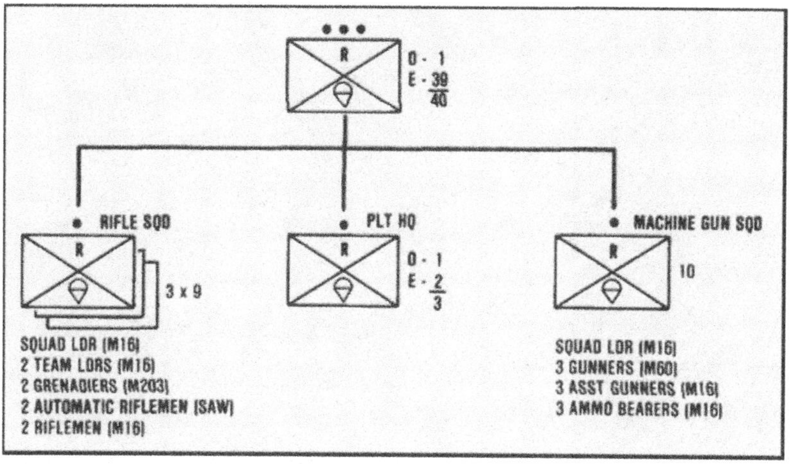

Figure A-10. Ranger rifle platoon.

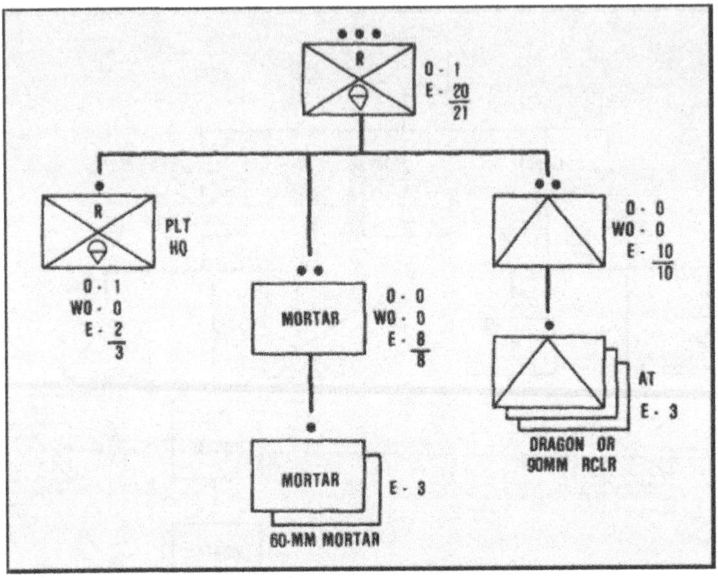

Figure A-11. Ranger weapons platoon.

APPENDIX B

Composition and Equipment of Ranger Regiment Liaison Teams

Team Red	RANK
Liaison Officer	MAJ
Senior Intelligence Analyst	SSG
Clerk-Typist	SP4
Tactical Communications System Operator/Mechanic	SGT
Combat Telecommunications Center Operator	SP4
Tactical Communications System Operator/Mechanic	SP4
Team White	
Liaison Officer	CPT
Senior Intelligence Analyst	SSG
Clerk-Typist	SP4
Tactical Communications System Operator/Mechanic	SGT
Combat Telecommunications Center Operator	SP4
Tactical Communications System Operator/Mechanic	SP4

Equipment (for Each Team)

HMMWV with 3/4-ton trailer	5-kw generator
Communications shelter	Tent, GP, small
AN/PCS-3 radio set (3)	AN/VSC-7 net control station (1)
MX 300 (4)	KY 57 with battery case (5)
HYX 57 with battery case (4)	HYP 57 (5)

KYK 13 (2)

KY 65 with battery case (1)

KG 84 (1)

H189 (2)

GRC 193 (1)

VGC-74 (1)

PP 1648 (1)

SB-22 switchboard (1)

KYI 18 (1)

KY 65 WLA (1)

Digital message device (1)

AN/PRC-77 radio (1)

Facsimile transmission device (1)

TA 312 with WD1 (5)

APPENDIX C

Equipment Density

C-1. Communication equipment.

	Regimental Headquarters	Battalions (x3)
AN/VSC-7 net control station	7	2
AN/PSC-3 radio set	23	13
AN/PRC-41 radio set	2	5
AN/PRC-104 radio set	9	26
AN/GRC-193A radio set	4	0
AN/PRC-70 radio set	6	0
AN/PRC-77 radio set	22	50
AN/VRC-46 radio set	2	0
AN/VRC-49 radio set	0	2
AN/PRC-68 radio set	0	66
TSEC/KY-57 speech security device	60	68
TSEC/KY-58 speech security device	6	0
TSEC/KY-65 speech security device	11	0
AO-8990/P digital message device group	15	23
AN/GXC-7A tactical FAX device	7	2
AN/UGC-74A(V)3 communications terminal	7	0
SB-22 switchboard	2	0
SB-3614(V)/TT switchboard	2	1

TSEC/KG-84	5	0
SB-993/GT switchboard	0	12

C-2. Weapons and equipment each battalion.

Infrared aiming light	96	Interrogator set, Stinger missile	3
Rifle, M16	451	Grenade launcher, M203	67
Rocket launcher, M202	13	Machine gun, M60	27
Mortar, 60-mm M224	6	Rifle, sniper, M21	24
Recoilless rifle, 90-mm	18	Viewer, infrared	9
Tracker, Dragon	9	Night vision devices (all types)	322
Truck, 1/4-ton	2	Laser target designators	11
Radio (all types)	164	Tactical satellite radios (all types)	15
Squad automatic weapon	54		
Laser range finder	14		

APPENDIX D

Standing Orders, Rogers' Rangers

1. Don't forget nothing.
2. Have your musket clean as a whistle, hatchet scoured, sixty rounds powder and ball, and be ready to march at a minute's warning.
3. When you're on the march, act the way you would if you was sneaking up on a deer. See the enemy first.
4. Tell the truth about what you see and what you do. There is an army depending on us for correct information. You can lie all you please when you tell other folks about the Rangers, but don't never lie to a Ranger or officer.
5. Don't never take a chance you don't have to.
6. When we're on the march we march single file, far enough apart so one shot can't go through two men.
7. If we strike swamps, or soft ground, we spread out abreast, so it's hard to track us.
8. When we march, we keep moving till dark, so as to give the enemy the least possible chance at us.
9. When we camp, half the party stays awake while the other half sleeps.
10. If we take prisoners, we keep 'em separate till we have had time to examine them, so they can't cook up a story between 'em.
11. Don't ever march home the same way. Take a different route so you won't be ambushed.
12. No matter whether we travel in big parties or little ones, each party has to keep a scout 20 yards ahead, 20 yards on each flank, and 20 yards in the rear so the main body can't be surprised and wiped out.
13. Every night you'll be told where to meet if surrounded by a superior force.
14. Don't sit down to eat without posting sentries.
15. Don't sleep beyond dawn. Dawn's when the French and Indians attack.
16. Don't cross a river by a regular ford.

17. If somebody's trailing you, make a circle, come back onto your own tracks, and ambush the folks that aim to ambush you.

18. Don't stand up when the enemy's coming against you. Kneel down, lie down, hide behind a tree.

19. Let the enemy come till he's almost close enough to touch. Then let him have it and jump out and finish him up with your hatchet.

MAJ Robert Rogers

APPENDIX E

The Ranger Creed

R Recognizing that I volunteered as a Ranger, fully knowing the hazards of my chosen profession, I will always endeavor to uphold the prestige, honor, and high "esprit de corps" of the Ranger Regiment.

A Acknowledging the fact that a Ranger is a more elite soldier who arrives at the cutting edge of a battle by land, sea, or air, I accept the fact that as a Ranger, my country expects me to move farther, faster, and fight harder than any other soldier.

N Never shall I fail my comrades. I will always keep myself mentally alert, physically strong, and morally straight and I will shoulder more than my share of the task, whatever it may be. One hundred percent and then some.

G Gallantly will I show the world that I am a specially selected and well-trained soldier. My courtesy to superiors, my neatness of dress, and my care of equipment shall set the example for others to follow.

E Energetically will I meet the enemies of my country. I shall defeat them on the field of battle, for I am better trained and will fight with all my might. Surrender is not a Ranger word. I will never leave a fallen comrade to fall into the hands of the enemy and under no circumstances will I ever embarrass my country.

R Readily will I display the intestinal fortitude required to fight on to the Ranger objective and complete the mission, though I be the lone survivor.

RANGERS LEAD THE WAY

APPENDIX G

Training

G-1. Application and guidance.

a. To prepare for combat, ranger units train on missions derived from their analysis of operational and contingency plans and the ranger mission training plans (MTPs). Training guidance is in doctrinal manuals, field circulars, and soldier's training publications. Training must ensure that the ranger unit is skilled in all basic light infantry skills and in ranger-unique skills. In the final analysis, good training depends upon the leadership and teamwork of ranger unit officers and NCOs.

b. Everything rangers do should prepare them to perform well in combat. Ranger units train IAW the battalion training management system (BTMS) concentrating on basic skills and on live-fire, hands-on training.

c. An effective training program breeds high morale and few discipline-related problems. Some qualities of ranger units that have good training programs are as follows:

(1) Leaders who can solve problems.

(2) A thorough and continuous analysis of the unit's mission.

(3) Validated training objectives.

(4) Attention to the basics (move, shoot, and communicate).

(5) Multiechelon sustainment training.

(6) Standardized procedures.

(7) Physical and mental strength.

(8) Live fire as an integral part of training.

(9) Safety as a basis to all operations.

(10) The use of night and adverse weather and difficult terrain as combat force multipliers,

d. When all training is planned with realism, rangers will be in training situations that require the same attention as in battle. Realism and stress in training result in faster learning and better retention of skills. Training objectives must be clear with measurable standards to determine proficiency. Rangers and ranger units must train to the standards that will be expected in combat.

e. Examples of unit training goals include:

(1) Developing each member of the chain of command to perform his duties with competence, confidence, and pride.

(2) Maintaining small unit integrity during all training and support activities. Accomplishing tasks as units.

(3) Ensuring that all tasks for training are keyed to mission needs and weaknesses. Assets--ranger's time, money, and materiel--are expended only after thorough planning.

(4) Training to standard.

(5) Developing and maintaining discipline and esprit to ensure the abilities of each ranger and all units are achieved.

(6) Emphasizing use of the ranger mission training plan (MTP) for each echelon of the organization.

(7) Emphasizing the basics.

(8) Maintaining a high level of physical fitness.

(9) Working together within the ranger regiment and other elements of the special operations forces community.

(10) Developing a unit atmosphere of quick reaction under stress. Encouraging and rewarding individual initiative. Accomplishing unit missions IAW the commander's intent.

G-2. Philosophy.

a. To be effective, training should be physically and mentally challenging. Each ranger will then have a sense of purpose, accomplishment, and satisfaction. Training is evaluated by the results rather than hours of training. Units that are well trained today need reinforcement training to sustain readiness. Training management must be continuous. Training managers must use the time available to improve training efficiency. Cyclic training will not produce the desired results. The "train-to-correct deficiencies" method is needed. The training managers must analyze weaknesses, conduct prescriptive and reinforcement training, and reevaluate their unit's training status. The process that is used to develop and maintain a continuous training and evaluation program is shown in Figure G-1.

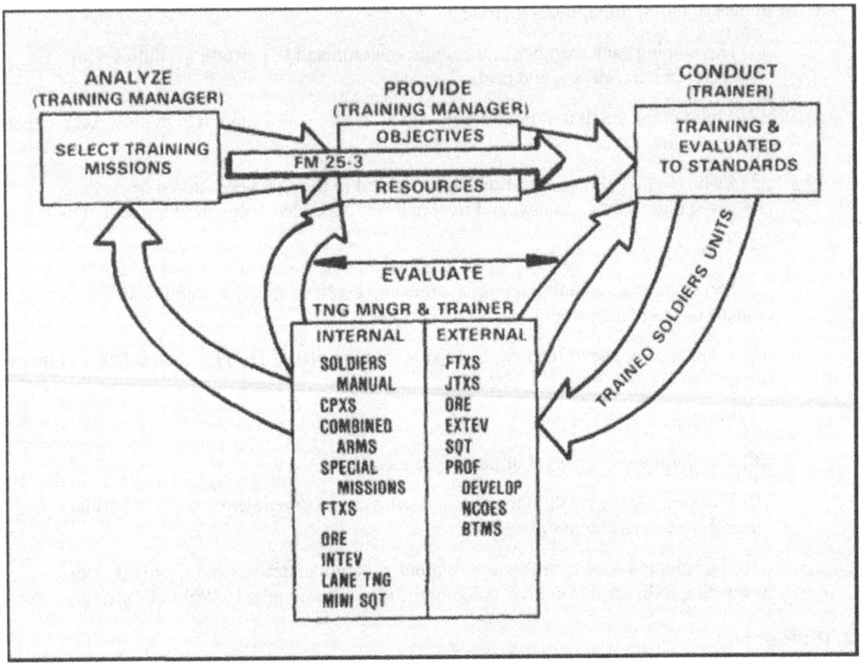

Figure G-1. The Army training and evaluation program.

b. Individual and collective training must be integrated within a unit training program. Training of the individual ranger, leaders, and units at each level must occur at the same time. The ranger commander or leader plans and conducts training for every member of his unit.

c. Ranger commanders analyze operational and contingency plans to determine unit missions and translate the associated tasks into training objectives. These objectives form the basis of the unit training program. The regimental headquarters provides training guidance in a formal regimental training plan. Commanders at lower levels program, supervise, and conduct training within their units.

d. All members of the chain of command are responsible for training. The NCO leader or first-line supervisor conducts individual and small unit collective task training. The officer plans individual training in his unit and directs and conducts collective training.

e. Centralized development of training objectives and decentralized training builds well-trained units and outstanding leaders. Authority and responsibility for the detailed planning, scheduling, conducting, evaluating, and supervising of training are delegated to the lowest element capable of managing this training. Higher headquarters is responsible for establishing objectives, monitoring and evaluating training, and issuing general guidelines.

f. Multiechelon training and evaluation means using all assets and time available to improve combat readiness. The multiechelon structure of the ranger MTP provides a basis for units to be continually trained and evaluated by internal evaluation. Each leader must know his mission tasks and train to accomplish those tasks to the set standards.

G-3. Mission analysis.

a. Mission analysis is continuous for a good training program. The selection of training missions or tasks results from analysis of the unit's overall mission. (See Figure G-1 above.) Tasks and subtasks are then placed in order of priority.

b. The trainer must identify countertasks for the opposing force (OPFOR). The countertasks should be such that an OPFOR that meets the standards for its countertask will be successful. Both sides have a chance to win.

c. The trainer should then develop combat drills and situational training exercises (STXs), if not furnished by the proponent school.

G-4. Unit training.

a. **Responsibility.** Ranger commanders at all levels are responsible for all training and management described herein and current Army training doctrine. This includes individual and collective training at all levels. Units must train to the standards specified in training objectives.

b. **Management.** The training management model for commanders is:

- Set objectives.
- Provide resources.
- Coach subordinates.
- Measure results.
- Provide feedback.

G-5. Individual training.

There are three elements basic to a well-trained individual ranger. He must be physically fit, be proficient in the use of his weapon, and have the psychological commitment to fight the enemy.

a. Individual training is the responsibility of the first-line supervisor under the supervision of the chain of command. It is conducted using performance-oriented techniques under battlefield conditions. New skills must be developed and practiced, and acquired skills must be reinforced throughout the training year. A high level of proficiency must be maintained through continuous training and practice.

b. Trainers should use peer instruction. Commanders ensure needed resources are available to conduct challenging individual training. Ranger units schedule and conduct this training and evaluate individual proficiency.

c. Flexibility in planning and scheduling is necessary. The goal of all individual training must provide for reinforcement training to offset the effects of the forgetting curve. (See Figure G-2).

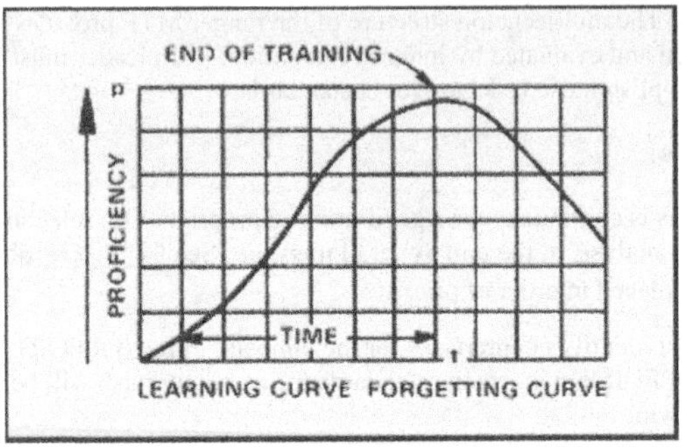

Figure G-2. The learning curve and forgetting curve.

d. The command sergeants major, unit first sergeants, and staff section sergeants are the keys to an effective training program. These senior NCOs must counsel, guide, and critique junior NCOs in training preparation and execution. Fire team leaders, section chiefs, and squad leaders coordinate and present individual training.

e. The NCO supervisors must ensure that the ranger is trained in the skills required for the specific military occupational specialty (MOS) and in those skills used daily. Not all training can be acquired in an office. For example, all rangers must maintain a high skill level with their weapons, be trained in NBC subjects, and be capable of performing contingency missions.

G-6. Collective training.

a. **Guidance.** Collective training trains crews, teams, squads, platoons, companies, and battalions to do their missions. The ranger MTP is used to plan training and evaluate the unit's weaknesses and strengths. If a ranger commander receives new training requirements or missions, they should become part of the collective training program. The ranger MTP describes the combat tasks to be performed by every echelon of the ranger battalion. It is a plan for organizing, conducting, and evaluating the year-round unit training program.

b. **Responsibility.** The ranger unit commander is responsible for the conduct of collective training and for all other training in his unit. Practical exercises emphasize the techniques of terrain use, movement, and effective employment of weapon systems. Training under stress will be conducted to move the unit to a higher plane of combat readiness.

c. **Drills.** Drills are written for small units--for example, fire team and squad, crew and section--and describe an action that is shorter than an entire STX. Drill training objectives cover both individual and collective skills (critical collective tasks) and list them together for effective group action. Drills must be mastered in order to react to rapidly changing situations in combat. The trainer can interrupt the execution of a drill to provide necessary training on individual or collective skills.

d. **Situational training exercise.**

(1) The STXs are short, mission-oriented tactical exercises in which a group of closely related collective tasks are trained. They are like drills but are more complex and flexible. They usually involve a larger unit than does a drill. They integrate leader training, leader tasks, drills, and separate individual tasks into a realistic setting. Trainers may execute STXs as written or tailor them to meet specific unit requirements. An STX is not standardized, but it must have a firm doctrinal base.

(2) Due to their design and purpose, STXs apply mainly to platoon- and company-size units. This does not preclude the development of STXs or similar exercises for unit staffs and headquarters. They combine "how-to-train" specifics for selected collective tasks with the "what-to-train" specifics found in doctrinal manuals and other documents.

G-7. Performance-oriented training.

a. Performance-oriented training emphasizes hands-on performance. Training is conducted to ensure a level of proficiency, rather than established number of hours of required training.

b. Performance-oriented training has precise training objectives and makes good use of resources. Training objectives serve as the basis for preparing, conducting, and evaluating all training.

G-8. Evaluation.

a. A training program must include an evaluation of training (planned and conducted). An evaluation plan must:

(1) Provide immediate feedback to the trainer.

(2) Recognize effective trainers.

(3) Evaluate all units in the conduct of training to provide for a comparison of training programs.

(4) Be flexible to unit needs; provide general and specific evaluations based on the level of training within the unit.

(5) Be in effect at all times.

b. Major considerations for training evaluations include:

(1) Ranger unit commanders establish and maintain an effective internal and external evaluation program.

(2) The director of each staff element is responsible for training evaluations within his area of staff responsibility with emphasis on low-density MOS groups.

(3) The operations officer of the ranger organization develops and validates training and evaluation plans.

c. Evaluation of training must be done vertically from the major task (mission) down through subtasks (collective tasks) to the leader and individual tasks. In selection of individual tasks for evaluation, priority should be given to critical individual tasks that impact on the success of the collective task or mission.

d. The critique and after-action review, after each major training event, are the most important aspects of training. The after-action review takes place immediately after completion of a training task (individual or collective). The individual(s) self-analyzes and critiques the training event. Then after a formal critique of the training, the individuals) trains to standard (not to time).

G-9. Field training.

a. **Considerations.** Ranger commanders schedule training under field conditions to permit maximum training opportunities. Field training emphasis is on mastery of a series of individual (soldier and leader) tasks and collective training tasks. These tasks are repeated until established standards are met. The goal of ranger field training is to ensure realistic, demanding, and meaningful training.

(1) When training is conducted under conditions of moderate stress, most combat critical skills are learned quicker and retained longer. (See Figure G-3 on the following page.) Stress can be incorporated in a training situation by physical deprivation (sleep, food), by live-fire exercises, by operation against an OPFOR under simulated battlefield conditions for an extended period, by the intensity of activity including demands for decision making, or by physically demanding activities. Ranger units train under stress so that they will react quickly in combat.

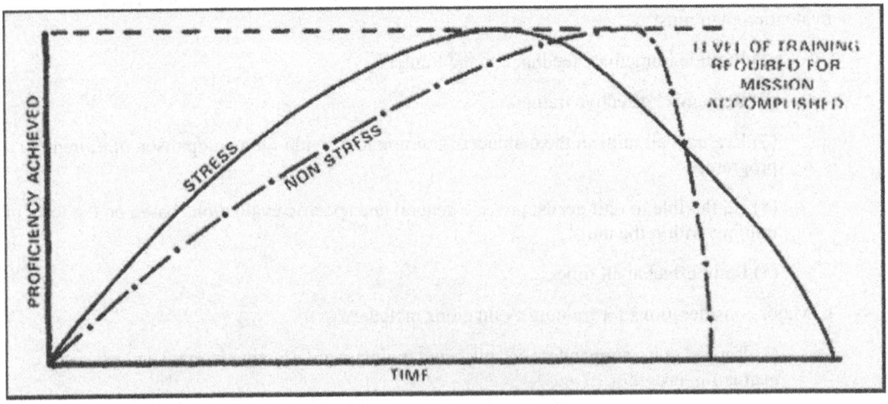

Figure G-3. The Army training and evaluation program.

(2) Antagonistic training situations are two-sided exercises conducted by opposing elements. Each element has a training objective (task and standards). The opposing force has a corresponding countertask. This training may be progressive--for example, starting with a

crawl-through, then a walk-through, and progressing to both elements operating at a fast pace. The use of multiple integrated laser equipment system (MILES) greatly enhances task and countertask training. The units should also be considered for a live-fire exercise against a realistic OPFOR target array when the basics have been mastered.

b. **Methods of training.**

(1) **Considerations.** The chance for success in combat can be increased by becoming proficient in training situations that depict realistic combat conditions. (See Figure G-4.) Meeting the standards of a realistic live-fire training exercise assures the commander that his unit could perform well in combat. Figure G-4 shows the relationship between realistic training and the challenge to the trainer.

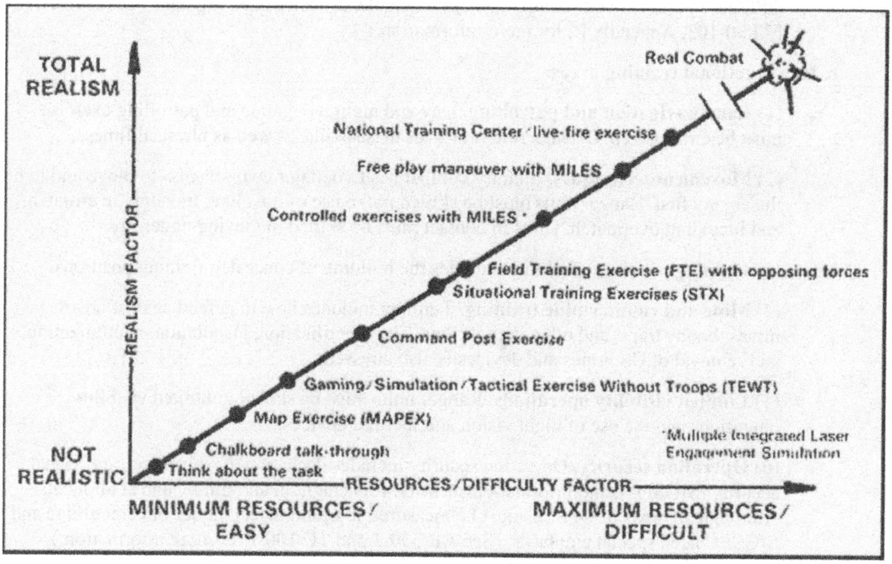

Figure G-4. Training proficiency scale.

(2) **Ranger cohesiveness training.** Ranger units promote teamwork by undergoing tough, stressful, often hazardous training. This challenges the unit, which gains in cohesiveness when it meets these challenges. This includes serving together in live-fire exercises, athletic events, precision parades, high-standard inspections, tough physical training, and combat.

(3) **Field training exercise.** An FTX is training normally conducted under simulated, realistic combat conditions. This training normally exercises command and control systems and the administrative and logistical needs of sustained combat. Stress training is part of this exercise, focusing on live-fire and MILES drills and STXS.

(4) **Combined arms training.** Combined arms training should include all means of support such as tactical air, armed helicopters, and indirect fire with emphasis on fire support coordination and airspace control.

(5) **Live-fire exercises.** Including live-fire exercises in drills and STXs offers a chance to evaluate units in their response to simulated combat missions. Ranger units train under pressure and always anticipate being selected to conduct a live-fire exercise. Live-fire exercises develop confidence and esprit, and reinforce unit discipline.

(6) **Opposing force.** Committing one unit against another adds a competitive spirit to a field training exercise (FTX). The use of MILES and squad combat operations exercises (simulation) (SCOPES) provides realism to two-sided field exercises. Trainers should use OPFOR equipment and training devices available, as well as making their own devices. (See FM 30-103, Appendix R, for more information.)

c. **Key functional training areas.**

(1) **Land navigation and patrolling.** Day and night navigation and patrolling exercises must be emphasized. Orienteering improves these skills, as well as physical fitness.

(2) **Movement techniques.** Infantry combat has two major components--to move and to hit the enemy first. Ranger units must be skilled in the use of traveling, traveling overmatch, and bounding overmatch. Units in contact must be skilled in moving under fire.

(3) **Fighting** positions. Training includes the building of concealed fighting positions.

(4) **Mine and countermine training.** Training includes how to defend against hostile mines, booby traps, and other demolitions. The identification, installation, neutralization, and removal of US mines and devices is also stressed.

(5) **Limited visibility operations.** Ranger units must be skilled in limited visibility operations and the use of night vision and thermal devices.

(6) **Operation security.** Operation security includes deception, physical security, signal security (SIGSEC), and information security. It is taught in all training and at all levels. Although all areas of OPSEC must be measured in training, techniques of camouflage and SIGSEC need special emphasis. (See AR 530-1 and TC 100-1 for more information.)

G-10. Professional development.

Ranger units at battalion and regimental levels conduct many officer and NCO professional development programs. This ensures that all officers and NCOs are skilled in their field, confident, and wellrounded. Professional readings and book reports, tactics seminars, and tactical exercises without troops are good development methods.

G-11. Physical training.

a. This program ensures that each ranger maintains the physical ability and stamina needed to do his mission. Other objectives are to promote unit cohesiveness and discipline. These are met by combining individual testing, formation, running, athletic competition, weight control, proper diet, teamwork drills, and freedom from drug and alcohol abuse. Mental fitness and toughness come from experience and success. Good physical conditioning increases a unit's chances of success.

b. Physical training (PT) programs are designed to include the elements of overload, progression, balance, variety, and regularity.

c. Ranger unit physical fitness standards are:

(1) Score 80 points for each Army physical readiness test (APRT) event and do six chinups.

(2) Pass the ranger swimming test.

(3) Complete an 8-kilometer run in 40 minutes.

(4) Complete a 12-kilometer road march in three hours (with rucksack, helmet, and weapon).

(5) Meet the Army height and weight standards.

G-12. Environmental and specialized training.

a. Rangers conduct training in environments as close to real combat missions as possible. This conditions rangers to survive and fight in a specific environment. Such training includes combat-critical tasks and technique training for that environment.

b. Each ranger battalion's environmental and specialized training includes at least the following:

(1) Jungle (once every 12 months).

(2) Desert (once every 12 months).

(3) Urban (once every 6 months).

(4) Extreme cold (once every 18 months).

(5) Mountain (once every 12 months).

(6) Amphibious (once every 18 months).

G-13. Ranger indoctrination program.

a. The RIP is a training and selection program. It is run by specially selected NCOs from the regimental headquarters. It has as its mission the training and indoctrination of newly assigned soldiers. It physically toughens and mentally sharpens the newly assigned soldier. It teaches basic skills and techniques. The RIP identifies and eliminates in three weeks any applicant who does not show dedication, motivation, physical fitness, and emotional stability.

b. An indoctrination program for newly assigned ranger-qualified officers and NCOs prepares them to assume duty as leaders within the ranger regiment. These personnel monitor and participate in RIP training, as well as undergoing individual ranger leadership training. The RIP provides them an overview of the regiment's history, its mission, and its goals. (See Appendixes C,

D, and E for further information about the history and goals of the ranger regiment.) The RIP also provides a review of the skills required in a ranger unit. Refresher training in airborne and pathfinder operations, use of indirect fires, and light infantry doctrine is conducted. During RIP all newly assigned officers and NCOs must meet the ranger physical fitness standard.

c. The RIP cadre also conducts a preranger program to prepare soldiers to attend the US Army Ranger School. It is a physically demanding course that teaches troop-leading procedures, operation orders, and small unit patrol actions.

Glossary

The following acronyms, abbreviations, and definitions are provided to aid the reader in understanding this publication.

Section I

ACRONYMS AND ABBREVIATIONS

AAT	Army assault team
ABCCC	airborne battlefield command and control center
ACL	allowable cargo load
ADA	air defense artillery
AFFOR	the USAF element of a joint task force
AGL	above ground level
AKZ	armor kill zone
ALCC	airlift control center
ALCE	airlift control element
AM	amplitude modulation
AMC	air mission commander
APERS	antipersonnel
APRT	Army physical readiness test
ARFOR	the Army element of a joint task force
ARSOC	Army special operations command
ARSOF	Army Special Operations Forces
ASG	area support group
ASPC	all-source production center

ATGM	antitank guided missile
AWADS	adverse weather aerial delivery system
BCE	battle coordination element
BTMS	battalion training management system
CA	civil affairs
CAS	close air support
CCT	combat control team
CDS	container delivery system
CE	communications-electronics
CI	counterintelligence
CINC	commander in chief
COMINT	communications intelligence
COMSEC	communications security
CONUS	continental United States
COSCOM	corps support command
CP	command post
CS	combat support; a chemical agent
CSS	combat service support
DACG	departure airfield control group
DACO	departure airfield control officer
DASC	direct air support center
DMDG	digital message device group
DOD	Department of Defense
DS	direct support
DSU	direct support unit
DZ	drop zone
EAC	echelons above corps

EACIC	Echelons Above Corps Intelligence Center
ECCM	electronic counter-countermeasure
ECM	electronic countermeasures
E&E	escape and evasion
EEFI	essential elements of friendly information
EOC	emergency operations center
EW	electronic warfare
FARP	forward area rearm and refuel point
FASCAM	family of scatterable mines
FAX	facsimile transmission
FDC	fire direction center
FEBA	forward edge of the battle area
FFL	friendly front lines
FIST	fire support team
FLOT	forward line of own troops
FM	field manual; frequency modulation
FO	forward observer
FORSCOM	US Army Forces Command
FPF	final protective fire
FPL	final protective line
FRAGO	fragmentary order
FSA	fire support area
FSCOORD	fire support coordination or coordinator
FSE	fire support element
FSO	fire support officer
FSS	fire support station
FTX	field training exercise

G3	Assistant Chief of Staff, Operations
GS	general support
GSR	ground surveillance radar
HAARS	high-altitude, airdrop resupply system
HAHO	high-altitude, high-opening
HALO	high-altitude, low-opening
HE	high explosive
HF	high frequency
HHC	headquarters and headquarters company
HSLLADS	high-speed, low-level airdrop system
HUMINT	human intelligence
IAW	in accordance with
IFF	identification-friend-or-foe
IP	initial point
IRP	initial rally point
ISB	intermediate staging base
JACC/CP	joint airborne communications center/command post
JCS	Joint Chiefs of Staff
JCSE	joint communications support element
JFC	joint force commander
JSOC	joint special operations command
JTB	joint transportation board
JTF	joint task force
kw	kilowatt
LAPES	low altitude parachute extraction system
LAW	light antitank weapon
LNO	liaison officer

LOC	logistical operations center
LRRP	long-range reconnaissance patrol
LRSU	long-range surveillance unit
LTD	laser target designator
LZ	landing zone
MAC	Military Airlift Command
MANPADS	man-portable air defense system
MCC	movement control center
METT-T	mission, enemy, terrain, troops and time available
MI	military intelligence
MILES	multiple integrated laser engagement system
mm	millimeter
MMC	Materiel Management Center
MMEE	minimum mission-essential equipment
MOS	military occupational specialty
MOUT	military operations on urbanized terrain
MRLS	multiple rocket launch system
MTP	mission training plan
NBC	nuclear, biological, chemical
NCA	National Command Authority
NCO	noncommissioned officer
NOD	night observation device
OCOKA	observation, concealment, obstacles, key terrain, and avenues of approach
OCONUS	outside continental United States
OPCOM	operational command
OPCON	operational control
OPFOR	opposing force

OPORD	operation order
OPSEC	operations security
ORP	objective rally point
PAC	personnel administration center
PDF	primary direction of fire
PIR	priority intelligence requirements
PLL	prescribed load list
POM	preparation for oversea movement
PSYOP	psychological operations
PT	physical training
PZ	pickup zone
QRE	quick-reaction element
RATELO	radiotelephone operator
REMAB	remote marshalling base
RFL	restrictive fire line
RIP	ranger indoctrination program
RP	rally point; release point
RPG	rocket-propelled grenade
RRF	ranger ready force
RRP	reentry rally point
RSE	ranger support element
RSTA	reconnaissance, surveillance, and target acquisition
S1	adjutant
S2	intelligence officer
S3	operations and training officer
S4	supply officer
S5	civil affairs officer

SAR	search and rescue
SATCOM	satellite communications)
SCIF	sensitive compartmented information facility
SCOPES	squad combat operations' exercises (simulation)
SCUBA	self-contained underwater breathing apparatus
SEAD	suppression of enemy air defense
SEMA	special electronic mission aircraft
SERE	survival, evasion, resistance, and escape
SF	special forces
SHORAD	short-range air defense
SIGSEC	signal security
SKE	station-keeping equipment
SLAR	side-looking airborne radar
SOA	special operations aviation
SOCB	special operations communications battalion
SOCOM	special operations command (usually, 1st SOCOM)
SOP	standing operating procedures
SOSB	special operations support battalion
SOTF	special operations task force
SPL	support platoon leader
STX	situational training exercise
TA	theater Army
TAACOM	Theater Army Area Command
TAC	Tactical Air Command
TACP	tactical air control party
TACSAT	tactical satellite terminal
TAMMC	Theater Army Materiel Management Center

TASG	theater Army support group
TDA	table of distribution and allowances
TOA	transfer of authority
TOC	tactical operations center
TOE	table(s) of organization and equipment
TRP	target reference point
TTY	teletype
UHF	ultrahigh frequency
US	United States
USAF	United States Air Force
USMC	United States Marine Corps
USN	United States Navy
VHF	very high frequency
WP	white phosphorus
XO	executive officer

Section II

DEFINITIONS

ADVANCE FORCE. An operation requiring the immediate deployment of US forces with the intent to negate a threat before the deployment of follow-on forces is required.

AIRLAND BATTLE DOCTRINE. An approach to military operations that realizes the full potential of US forces. Two notions (extending the battlefield and the ability to integrate conventional, nuclear, chemical, and electronics means) are combined to describe a battlefield where the enemy is attacked to the full depth of his formation.

AREA OF INFLUENCE. A geographical area wherein a commander is directly capable of influencing operations by maneuver or fire support systems normally under his command or control.

AREA OF INTEREST. That area of concern to the commander, including the area of influence, areas adjacent thereto, and extending into enemy territory to the objectives of current or planned operations. This area also includes areas occupied by enemy forces who would jeopardize the accomplishment of the mission.

AREA OF OPERATIONS. That portion of an area of conflict necessary for military operations. Areas of operations are geographical areas assigned to commanders for which they have responsibility and in which they have authority to conduct military operations.

CHALK. One specific aircraft load.

CLANDESTINE OPERATIONS. Activities to accomplish intelligence, counterintelligence, and other similar activities sponsored or conducted by governmental departments or agencies, in such a way as to ensure secrecy or concealment. (They differ from covert operations in that emphasis is placed on concealment of the operation rather than on concealment of the identity of the sponsor.)

COMMUNICATIONS INTELLIGENCE (COMINT). The interception and processing of foreign communications passed by radio, wire, or other electromagnetic means, and by the processing of foreign encrypted communications, however transmitted. Interception comprises search, intercept, operator identification, signal analysis, traffic analysis, crypto-analysis, study of plain text, the fusion of these processes, and reporting the results.

COMMUNICATIONS SECURITY. The protection resulting from all measures designed to deny to unauthorized persons information of value that might be derived from the possession and study of telecommunications, or to mislead unauthorized persons in their interpretations of the results of such study. COMSEC includes crypto-security, physical security, and transmission security.

COUNTERINTELLIGENCE. That aspect of intelligence activities, both offensive and defensive, designed to detect and neutralize or destroy the effectiveness of hostile foreign intelligence activities and to protect information against espionage, personnel against subversion, and installations and materiel against sabotage.

COVERT OPERATIONS. Operations that are so planned and executed as to conceal the identity of or permit plausible denial by the sponsor. They differ from clandestine operations in that emphasis is placed on concealment of identity of the sponsor rather than on concealment of the operation.

DECEPTION. Those measures designed to mislead the enemy by manipulation, distortion, or falsification of evidence to induce him to react in a manner prejudicial to his interests.

DEMONSTRATION. A show of force in an area where a decision is not sought. It is similar to a feint but does not make contact with the enemy.

DRILL. A drill focuses on a "chunk" or "slice" of battle and is targeted at small units. Drills are critical collective tasks that require a high degree of proficiency from a small unit. Generally, small units "drill" collective tasks that require rapid responses by unit members in the absence of detailed orders from unit leaders.

ELECTRONIC COUNTERMEASURES. That division of electronic warfare involving actions taken to prevent or reduce an enemy's effective use of the electromagnetic spectrum.

ELECTRONIC INTELLIGENCE (ELINT). The technical and intelligence information derived from foreign noncommunications electromagnetic radiations emanating from other than atomic detonations or radioactive sources.

ELECTRONIC WARFARE. Military action involving the use of electromagnetic energy to determine, exploit, reduce, or prevent hostile use of the electromagnetic spectrum and action that retains friendly use of the electromagnetic spectrum.

ELECTRONIC WARFARE SUPPORT MEASURES (ESM). That division of EW involving actions taken to search for, intercept, identify, and locate sources of radiated electromagnetic energy for the purpose of immediate threat recognition. ESM provides a source of information needed for immediate action involving electronic countermeasures, electronic counter-countermeasures, avoidance, targeting, and other tactical employment of forces.

ESPIONAGE. Actions directed toward the acquisition of information through clandestine operations.

EVACUATION/SAFEGUARD. The use of US forces to evacuate US citizens or allies from a hostile or potentially hostile area, or to safeguard US personnel or property.

EXFILTRATION. The removal of personnel or units from areas under enemy control by stealth, deception, surprise, or clandestine means. Exfiltration is a method of extracting ranger units.

EXTRACTION. The removal of personnel or units from areas under enemy control by any means, to include exfiltration, airland, airmobile, or amphibious.

FEINT. A diversionary supporting attack conducted to draw the enemy's attention from the main effort. It is normally executed by brigades and small units. Feints are usually shallow, limited objective attacks that go in before or during the main attack.

GUERRILLA WARFARE. Military and paramilitary operations conducted in enemy-held or hostile territory by irregular and predominantly indigenous forces.

HIGH-INTENSITY CONFLICT. A war between the US and any other nation or nations in which both belligerents use the most modern technology and all resources in intelligence, mobility, firepower (including NBC), command, communications and control, and service support.

HUMAN INTELLIGENCE. Intelligence derived from information provided by human resources.

INFILTRATION. The movement through or into an area of territory occupied by either friendly or enemy troops or organizations. The movement is made either by small groups or by individuals, at extended or irregular intervals. When used in conjunction with the enemy, it implies that contact is avoided. Infiltration is a subset of insertion.

INSERTION. Placement of troops and equipment into an operational area by any means.

INTERDICT. To isolate or seal off an area by military means; to prevent, hinder, or delay the use of an area or route by enemy forces.

INTERDICT LINE OF COMMUNICATIONS. An attack to seal off an area or to deny use of a route or approach.

JOINT TASK FORCE. A JTF may be constituted and designated by the Secretary of Defense or by the commander of a unified command, specified command, or an existing JTF. Normally, it performs missions having specific, limited objectives or missions of short duration. It dissolves when it has achieved its purpose. The joint task force commander is responsible to the commander, known as the establishing authority, who created the JTF. The JTF is composed of elements of two or more services operating under a single JTF commander. The JTF commander has operational control over the entire force and may have direct command of his own staff with representatives from the other services. He exercises logistics coordination or control only as necessary to meet his subordinate commanders' logistics needs.

LIAISON. That personal contact or communication maintained between elements of military forces to ensure mutual understanding and unity of purpose and effort.

LINE OF COMMUNICATIONS. All routes (land, water, and air) that connect an operating military force with one or more bases of operations and along which supplies and military forces move.

LOW-INTENSITY CONFLICT. Limited politico-military operations conducted to achieve political, social, economic, or psychological objectives. It is generally confined to a specific geographic area and is often characterized by operational constraints imposed on ranger units by political and economic considerations. It may involve intermittent combat against regular threat military forces, but it more often involves conflict with irregular, security, or paramilitary forces of a country or group involved in an operation that is counter to the best interests and national security of the United States. It is characterized by relatively low levels of combat and constraints placed on the weaponry, tactics, and movements of the ranger force. It may involve counterterrorist or antiterrorist operations and generally occurs before a pro forma declaration of war.

MID-INTENSITY CONFLICT. War, declared or undeclared, between the US and the regular forces of one or more nations and their respective allies, if any, in which the belligerents use the most modern technology and all resources in intelligence, mobility, firepower (excluding nuclear, chemical, and biological weapons), command, communications and control, and support. They are used for limited objectives under definitive policy limitations as to the extent of destructive power that can be used or the extent of geographical area that might be involved.

MILITARY STRATEGY. Art and science of using the armed forces of a nation to secure the objectives of national policy by application of force or the threat of force. Military strategy sets the fundamental conditions for operations.

NATIONAL COMMAND AUTHORITY. This includes the President and Secretary of Defense or the duly deputized alternates or successors. Commonly referred to as NCA.

OPERATIONAL COMMAND/CONTROL. Operational command (OPCOM) and operational control in joint force terminology both refer to the authority exercised by joint commanders over subordinate service components. Those terms are not interchangeable in joint operations. OPCOM applies to the authority exercised by commanders of unified commands. OPCON is the authority that subordinate joint task force commanders exercise in the conduct of specific operations. The authority that military departments exercise over their respective components is commonly referred to as command less operational command.

OPERATIONAL LEVEL OF WAR. The operational level of war uses available military resources to attain strategic goals within a theater of war. It is the theory of larger unit operations. It also involves planning and conducting campaigns. Campaigns are sustained operations designed to defeat an enemy force in a specified space and time with simultaneous and sequential battles. The disposition of forces, selection of objectives, and actions taken to weaken or to outmaneuver the enemy all set the terms of the next battle and exploit tactical gains. They are all part of the operational level of war. In AirLand Battle doctrine, this level includes the marshalling of forces and logistical support, providing direction to ground and air maneuver, applying conventional and nuclear fires in depth, and employing unconventional and psychological warfare.

OPERATIONS SECURITY. Those measures designed to protect information concerning planned, ongoing, and completed operations from unauthorized disclosure. It includes all actions a command takes to deny the enemy information about friendly units and their operations.

ORDER OF BATTLE (OB). The identification, strength, command structure, and disposition of the personnel, units, and equipment of any military force.

RAID. Usually a small-scale operation, involving a swift penetration of hostile territory to secure information, confuse the enemy, or destroy his installations. It ends with a planned withdrawal upon completion of the assigned mission.

RANGER MISSION. To conduct strike, raid, or special light infantry operations. Execution of ranger missions is normally directed by the National Command Authority. Rangers also support operational-level commanders during operations of limited duration when allocated or assigned by the theater commander.

RECONNAISSANCE. A mission undertaken to obtain, by visual observation or other detection methods, information about the activities and resources of an enemy or potential enemy; or to secure data concerning the meteorological, hydrographical, or geographical characteristics of a particular area.

SEARCH AND RESCUE. The use of aircraft, surface craft, submarines, specialized rescue teams, and equipment to search for and rescue personnel in distress on land or at sea.

SEIZURE. As an operational purpose, the capture of a voluntarily restricted portion of an enemy-controlled territory. Capture of an isolated land mass such as an island is usually categorized as a seizure.

SEIZE KEY FACILITY. An attack on an installation with the intent of holding it for a short period. Normally, rangers would make the assault to seize and secure an installation for follow-on forces.

SERVICE COMPONENT. Each service component commander is responsible for recommending the proper use of his forces and for accomplishing operational tasks assigned by the joint commander. He is also responsible for his service in matters of internal administration and discipline; training in own service doctrine, techniques, and tactics; designation of specific units to meet joint requirements; logistics functions normal to the component; tactical employment of service component forces; and service intelligence matters.

SHOW OF FORCE. A mission carried out to demonstrate US resolve, whereby US forces are deployed to defuse a situation that may be harmful to US interests or national objectives.

SIGNAL INTELLIGENCE (SIGINT). A generic term that includes COMINT and ELINT. Activities concerned with the collection and processing of foreign electromagnetic emissions for the purpose of producing intelligence information.

SITUATIONAL TRAINING EXERCISE. An STX is a short, mission-oriented exercise in which a group of closely-related collective tasks are trained, using a doctrinally preferred method of executing these tasks to established standards. An STX is drill-like in nature, but it is more complex and flexible than a drill. It usually involves a larger unit than a drill.

SPECIAL OPERATIONS. Military operations conducted by specially trained, equipped, and organized DOD forces against strategic or tactical targets in pursuit of national, military, political, economic, or psychological objectives. Special operations may support conventional military operations, or they may occur independently when the use of conventional forces is either inappropriate or infeasible. Sensitive peacetime operations, except for training, are normally authorized by the NCA and conducted under the direction of the NCA or designated commander. Special operations may include unconventional warfare, counter-terrorist operations, collective security, psychological operations, and civil affairs measures.

SPECIAL OPERATIONS FORCES. All civil affairs, psychological operations, ranger (active component), and special forces units within the US Army, US Army Reserve, and National Guard.

SPECIAL OPERATIONS COMMAND. The major subordinate command to FORSCOM exercising command and control of all active component Army special operations forces not placed under the operational command of unified or specified commands.

STRATEGIC LEVEL OF WAR. Military strategy uses the armed forces of a nation to secure the objectives of national policy by applying force or the threat of force. Military strategy sets the fundamental conditions for operations.

STRIKE. An attack that is intended to inflict damage on, seize, or destroy an objective. Sometimes referred to as direct action missions.

SURVEILLANCE. The systematic observation of aerospace, surface or subsurface areas, places, persons, or things by visual, aural, electronic, photographic, or other means.

TACTICAL. Pertaining to the employment of units in combat.

TACTICAL LEVEL OF WAR. Tactics are the specific techniques smaller units use to win battles and engagements that support operational objectives. Tactics employ all available combat, combat support, and combat service support. Tactics involve the movement and positioning of forces on the battlefield in relation to the enemy, the provisioning of fire support, and the logistical support of force before, during, and following engagements with the enemy. At corps and division, operational and tactical levels are not clearly separable, but they are guided by the same principles. An operation designed to defeat any enemy force in an extended area does so through operational maneuver and a series of tactical actions.

TACTICAL COMBAT INTELLIGENCE. That knowledge of the enemy, weather, and geographical features needed by a commander in the planning and conduct of combat operations. It is derived from the interpretation of information on the enemy and the environment.

TARGET ANALYSIS. A detailed and systematic examination of processed intelligence to identify and locate targets.

UNIFIED COMMAND. A unified command is established and designated by direction of the President. The command is organized under a single unified CINC and has a broad, continuing mission. The CINC is responsible to the Secretary of Defense for accomplishing his broad and continuing mission and operates under the strategic and operational direction of the JCS. A unified command is composed of assigned forces of two or more services. These forces, as service components, are under the operational command of the unified commander. The CINC exercises operational command authority through service component commanders. When tasked by higher authority, he may establish a subordinate JTF to conduct specific missions. Under emergency conditions, the CINC has the authority to use all facilities and supplies of assigned forces to accomplish his mission.

References

REQUIRED PUBLICATIONS

Required publications are sources that users must read in order to understand or to comply with this publication.

Field Manuals (FM)

FM 7-8	The Infantry Platoon and Squad (Infantry, Airborne, Air Assault, Ranger)
FM 7-10	The Infantry Rifle Company (Infantry, Airborne, Air Assault, Ranger)
FM 7-20	The Infantry Battalion (Infantry, Airborne, and Air Assault)

RELATED PUBLICATIONS

Related publications are sources of additional information. They are not required in order to understand this publication.

Army Regulations (AR)

AR 350-1	Army Training

Field Manuals (FM)

FM 10-52-1	Commander's Handbook for Water Usage in Desert Operations
FM 10-501	Airdrop of Supplies and Equipment: Rigging Containers
FM 10-512	Airdrop of Supplies and Equipment: Rigging Typical Supply Loads
FM 10-542	Airdrop of Supplies and Equipment: Rigging Loads for Special Operations
FM 10-547	Airdrop of Supplies and Equipment: Rigging the High Speed Aerial Delivery Container CTU-2/A
FM 10-553	Airdrop of Supplies and Equipment: Rigging Ammunition
FM 12-3-4	Echelons Above Corps (EAC) Personnel and Administrative Doctrine

FM 14-6	Comptroller/Finance Services in Theaters of Operations
FM 19-15	Civil Disturbances
FM 20-11-1	Military Diving (Volume 1)
FM 20-11-2	Military Diving (Volume 2)
FM 21-20	Physical Training Program
FM 21-26	Map Reading
FM 21-75	Combat Skills of the Soldier
FM 21-76	Survival, Evasion, and Escape
FM 21-77A(S)	Joint Worldwide Evasion and Escape Manual
FM 22-9	Soldier Performance in Continuous Operations
FM 23-67	Machinegun, 7.62-mm, M60
FM 25-1	Training
FM 25-2	Unit Training Management
FM 25-3	Training in Units
FM 25-4	How to Conduct Training Exercises
FM 25-5	Training for Mobilization and War
FM 25-7	Training Ranges
FM 26-2	Management of Stress in Army Operations
FM 29-47	Main Supply and Service Company, Forward Supply and Service Detachments, Supply and Service Battalion, Airborne Division
FM 29-51	Division Supply and Field Service Operations
FM 30-5	Combat Intelligence
FM 30-17	Counterintelligence Operations
FM 31-18	Long-Range Reconnaissance Ranger Company
FM 31-19	Special Forces Military Free-Fall Parachuting
FM 1-24	Special Forces Air Operations

FM 31-25	Special Forces Waterborne Operations
FM 31-70	Basic Cold Weather Manual
FM 34-1	Intelligence and Electronic Warfare Operations
FM 34-60A	Counterintelligence Operations (U)
FM 55-9	Unit Air Movement Planning
FM 57-38	Pathfinder Operations
FM 57-220	Basic Parachuting Techniques and Training
FM 63-3J	Combat Service Support Operations--Corps
FM 63-4	Combat Service Support Operations--Theater Army Area Command
FM 63-5	Combat Service Support Operations--Theater Army
FM 90-3	Desert Operations
FM 90-4	Airmobile Operations
FM 90-5	Jungle Operations
FM 90-6	Mountain Operations
FM 90-8	Counterguerilla Operations
FM 90-10	Military Operations on Urbanized Terrain (MOUT)
FM 90-10-1	An Infantryman's Guide to Urban Combat
FM 100-5	Operations
FM 100-20	Low Intensity Conflict
FM 100-26	The Air-Ground Operations System
FM 101-5	Staff Organization and Operations
FM 101-10-1	Staff Officers' Field Manual: Organization, Technical and Logistical Data (U)

Firing Table (FT)

FT 60-P-1	Firing Tables for Mortar, 60-mm: M224, Firing Cartridge, HE, M720, Cartridge, HE, M49A4, Cartridge, TP, M5OA3, Cartridge, WP, M302AI, Cartridge, Illuminating, M83A3

Soldier's Training Publication (STP)

STP 7-11B24-SM	Soldier's Manual MOS 11B, Infantryman, Skill Level 2/3/4

Technical Manuals (TM)

TM 10-1670-267-12&P	Operator's and Organizational Maintenance Manual Including Repair Parts and Special Tools List for High Altitude Airdrop Resupply System: 500 Pound Capacity

Training Circulars (TC)

TC 23-14	Sniper Training and Employment
TC 23-90	Mortar Training
TC 24-1	Communications-Electronics Operation Instructions, the CEOI
TC 31-20-5(C)	Special Forces Communications (U)
TC 90-6-1	Military Mountaineering

Other Related Publications

King, Michael J. Dr., *Leavenworth Papers, No. 11*, "Rangers Selected Combat Operations in World War II," Combat Studies Institute, USACGSC, Fort Leavenworth, KS

PROJECTED PUBLICATIONS

Projected publications are sources of additional information that are scheduled for printing but are not yet available. Upon print, they will be distributed automatically via pinpoint distribution. They may not be obtained from the USA AG Publications Center until indexed in DA Pamphlet 310-1.

Field Manuals (FM)

FM 7-70	Light Infantry Squad/Platoon
FM 7-71	Light Infantry Company
FM 7-72	Light Infantry Battalion
FM 7-73	Light Infantry Brigade

FM 23-36 60-mm Mortar, M224

Made in the USA
Monee, IL
05 November 2026

67810972R00184